# II International (XV Regional) Scientific Conference on Technogenic Systems and Environmental Risk (TSER 2018)



IOP Conference Series: Materials Science and Engineering
Volume 487

Obninsk, Russia
19 - 20 April 2018

**Editors:**

**Alla Oudalova**
**Dmitry Samokhin**
**Olga Momot**
**Boris Synzyznys**
**Helen Sarapul'tzeva**
**Natalia Epshtein**
**Ludmila Komarova**

ISBN: 978-1-5108-8330-7
ISSN: 1757-8981

**Printed from e-media with permission by:**

Curran Associates, Inc.
57 Morehouse Lane
Red Hook, NY 12571

**Some format issues inherent in the e-media version may also appear in this print version.**



Printed by Curran Associates, Inc. (2019)

For permission requests, please contact the Institute of Physics at the address below.

Institute of Physics
Dirac House, Temple Back
Bristol BS1 6BE UK

Phone: 44 1 17 929 7481
Fax: 44 1 17 920 0979

techtracking@iop.org

**Additional copies of this publication are available from:**

Curran Associates, Inc.
57 Morehouse Lane
Red Hook, NY 12571 USA
Phone: 845-758-0400
Fax: 845-758-2633
Email: curran@proceedings.com
Web: www.proceedings.com

# Table of contents

## Volume 487

**The II International (XV Regional) Scientific Conference "Technogenic Systems and Environmental Risk"**

**19–20 April 2018, Kaluga Region, Russian Federation**

**Accepted papers received: 23 January 2019**
**Published online: 13 March 2019**

### Preface



### Papers

IOP Conf. Series: Materials Science and Engineering **487** (2019) 011001 doi:10.1088/1757-899X/487/1/011001

# The II International (XV Regional) Scientific Conference "Technogenic Systems and Environmental Risk"

## (TSER-2018)

**April 19-20, 2018**
**Obninsk, Kaluga Region, RUSSIAN FEDERATION**

Accepted, peer reviewed papers from the 2nd TSER-2018

Edited by

Alla OUDALOVA, Dmitry SAMOKHIN, Olga MOMOT, Boris SYNZYZNYS, Helen SARAPUL'TZEVA, Natalia EPSHTEIN, Ludmila KOMAROVA

National Research Nuclear University "MEPhI"
(Moscow Engineering Physics Institute)
Moscow – Russian Federation

The scientific conference "Technogenic Systems and Environmental Risk" is held annually from 2004 to 2017 at Obninsk Institute for Nuclear Power Engineering, NRNU MEPhI. The main participants of the conference are students, graduate students and young scientists, including representatives of research institutes, universities, scientific and industrial associations and enterprizes. Since 2017, the meeting has officially been recognized as an international conference.

The II International (XV Regional) Scientific Conference "Technogenic Systems and Environmental Risk" 2018 (TSER-2018) was held in Obninsk, Kaluga Region, Russian Federation on April 19-20 2018. It was organized by National Research Nuclear University "MEPhI" (NRNU MEPhI)

The conference addresses current scientific problems related to the study of man-made factors in the human environment, the evaluation of the effect of these factors on humans and natural systems, ensuring the ecological safety of the biosphere. The meeting contributes to raising the level of knowledge of young scientists and teachers, sharing experience in conducting research and teaching activities, establishing links between organizations, regions and countries.

The Proceedings consists of accepted papers that are categorized under six topics, namely:

1) Nuclear and Thermal Power Engineering,
2) Modern Ecological Technologies. Waste Management,
3) Technogenic and Environmental Risks. Radiation Safety,
4) Biotesting and Bioindication of the Environment,
5) Ecological and Pharmaceutical Chemistry,
6) Effects of Non-Ionizing and Ionizing Radiation: Ecological Issues.

IOP Conf. Series: Materials Science and Engineering **487** (2019) 011001 doi:10.1088/1757-899X/487/1/011001

In 2018, the conference attracted a record number of participants. The organizing committee received more than 240 applications for participation from research and educational institutions of Obninsk, Kaluga, Moscow and the Moscow region (Dimitrovgrad, Pushchino, Mytishchi), Bryansk, Volgodonsk, Voronezh, Irkutsk, Yoshkar-Ola, Kazan, Kirov, Kursk, Murmansk, Nizhny Novgorod, Novocherkassk, Orel, Perm, Rostov-on-Don, Samara, Saratov, Sergach, Simferopol, Syktyvkar, Tomsk, Cherepovets, and also Minsk, Mogilev, Gomel (Belarus), Kurchatov, Ust-Kamenogorsk (Kazakhstan) , Bishkek, Karakol (Kyrgyzstan), Kiev (Ukraine a) Baku (Azerbaijan), Alexandria (Egypt). About 170 participants attended the venue, and 105 oral presentetations were delivered (including 21 in the school section).

We sincerely hope the 2$^{nd}$ THER-2018 Conference would provide a good opportunity to share experiences and ideas as well as to establish new academic collaborations among participants leading to exchange program and joint research and publication for the benefit of al***l.***

**Chairman:**
Alla OUDALOVA, D.Sc.
Chairman
Instutute of Nuclear Physics and Engineering
National Research Nuclear University “MEPhI”
Moscow, Russian Federation

IOP Conf. Series: Materials Science and Engineering **487** (2019) 011001 doi:10.1088/1757-899X/487/1/011001

## LOCAL ORGANIZING COMMITTEE

**Chairman:**
Alla OUDALOVA, D.Sc. INPhE NRNU MEPhI, Russia

**Co-chairmen:**
Barbashina N.S., Ph.D., INPhE NRNU MEPhI, Russia
Leonova T.N., D.Sc., NRNU MEPhI, Russia

**Members:**
Komarova L.N., D.Sc., PhysBio NRNU MEPhI, Russia
Maskevich S.A., D.Sc., Professor, ISEI BSU, Belarus
Momot O.A., Ph.D., INPhE NRNU MEPhI, Russia
Samokhin D.S., Ph.D., CAE NRNU MEPhI, Russia
Sarapultseva E.I., D.Sc., Prof., PhysBio NRNU MEPhI, Russia
Synzynys B.I., D.Sc., Prof., INPhE NRNU MEPhI, Russia
Rasskazova M.M., Ph.D., PhysBio NRNU MEPhI, Russia
Terekhova A.M., INPhE NRNU MEPhI, Russia
Epstein N.B., D.Sc., PhysBio NRNU MEPhI, Russia

**Conference Secretary:**
Kiseleva L.A., INPhE NRNU MEPhI, Russia
Shevenkova K.N., INPhE NRNU MEPhI, Russia

IOP Conf. Series: Materials Science and Engineering **487** (2019) 011001 doi:10.1088/1757-899X/487/1/011001

## INTERNATIONAL PROGRAM COMMITTEE

IOP Conf. Series: Materials Science and Engineering **487** (2019) 011001 doi:10.1088/1757-899X/487/1/011001

Kolesov V.V., Ph.D., Assoc. Prof., INPhE NRNU MEPhI, Russia
Komarova L.N., D.Sc., Head of Depart., PhysBio NRNU MEPhI, Russia
Kozmin G.V., Ph.D., Leading Researcher, RIRAE, Russia
Lavrentyeva G.V., Ph.D., Assoc. Prof., KB BMSTU, Russia
Lukashenko S.N., Deputy Director, National Nuclear Center of the Republic of Kazakhstan, Kazakhstan
Maskevich S.A., D.Sc., Prof., Director, ISEI BSU, Belarus
Nakhabov A.V., Ph.D., Assoc. Prof., INPhE NRNU MEPhI, Russia
Oudalova A.A., D.Sc., Head of the Depart., INPhE NRNU MEPhI, Russia
Polyakova L.P., Ph.D., Assoc. Prof., INPhE NRNU MEPhI, Russia
Perov S.G., D.Sc., Leading Researcher, Res. Inst. of MT, Russia
Samokhin D.S., Ph.D., Deputy Director of the CAE NRNU MEPhI, Russia
Sarapultseva E.I., D.Sc., Prof., PhysBio NRNU MEPhI, Russia
Segeda T.A., Ph.D, Head of sub-department «New technologies and materials in nuclear industry», D.Serikbaev EKSTU, Republic of Kazakhstan
Synzynys B.I., D.Sc., Prof., INPhE NRNU MEPhI, Russia
Tiras H.P., Ph.D., Leading Researcher, Institute of Experimental and Theoretical Biophysics of the Russian Academy of Sciences; Deputy Director , Pushchino State Natural Science Institute, Russia
Ulyanenko L.N., D.Sc., Prof., A. Tsyb MRRC, Russia
Shershakov V.M., D.Sc., CEO of RPA «Typhoon», Russia
Shilina A.S., Ph.D., Head of Depart., PhysBio NRNU MEPhI, Russia
Shubina O. A., PhD, Scientific Secretary, RIRAE, Russia
Yatsalo B.I., D.Sc., Prof., ICIS NRNU MEPhI, Russia

IOP Conf. Series: Materials Science and Engineering **487** (2019) 011001 doi:10.1088/1757-899X/487/1/011001

## KEYNOTE SPEAKERS

S.A. Maskevich, Professor, Director, International Sakharov Environmental Institute of Belarusian State University, Minsk, Belarus, 'Radiation monitoring of the environment: innovative technologies and informatization'

Yu.A. Kazanskiy, Professor, Obninsk Institute for Nuclear Power Engineering, NRNU MEPhI, Obninsk, Russia, 'Is the transmutation of heavy nuclei and fission fragments worthwhile?'

V.M. Shershakov, Professor, Director, RPA "Typhoon" of Roshydromet, Obninsk, Russia, 'Some results of determining the source and reasons for the appearance of 106Ru on the territory of Russia in September-October 2017'

IOP Conf. Series: Materials Science and Engineering **487** (2019) 011002 doi:10.1088/1757-899X/487/1/011002

# Peer review statement

All papers published in this volume of *IOP Conference Series: Materials Science and Engineering* have been peer reviewed through processes administered by the proceedings Editors. Reviews were conducted by expert referees to the professional and scientific standards expected of a proceedings journal published by IOP Publishing.

IOP Conf. Series: Materials Science and Engineering **487** (2019) 012001 doi:10.1088/1757-899X/487/1/012001

# Radiation monitoring of the environment: innovative technologies and informatization

**S A Maskevich**[1] and **M G Germenchuk**[2]

[1] International Sakharov Environmental Institute of Belarusian State University, Minsk, Belarus
[2] Scientific Engineering Center of Nuclear and Radiation Safety, Minsk, Belarus

E-mail: margermen@gmail.com

**Abstract.** Now the Republic of Belarus has a modern system of radiation monitoring of the environment, which is based on innovative technologies, including information technologies. The Republic of Belarus has good practices in organizing radiation monitoring of the environment after the accident at the Chernobyl NPP. Between 1986 and 2017, over 630,000 soil samples were measured for radionuclide contamination, including cesium-137. With the use of GIS technologies maps of radioactive contamination of the territory of Belarus with cesium-137, strontium-90, isotopes of plutonium and americium-241 have been created. New methodical approaches to the formation of a radiation monitoring system were developed in case of new threats to radiation safety, for example, the construction of nuclear power plants. Based on the experience of studying the behavior of radionuclides, an integrated biospheric approach to the organization of radiation monitoring of the environment is developed.

## 1. Introduction

Radioactive contamination is one of the most dangerous types of environmental contamination, both in terms of the negative impact on public health, and in terms of the subsequent inclusion of artificial radionuclides in biogeochemical natural cycles and their subsequent long-term effects on the elements of the biosphere.

Now in the Republic of Belarus there is a complicated radioecological situation, which is caused by global radioactive contamination due to nuclear tests of the environment and the consequences of the Chernobyl disaster. Currently, Belarus is building a Belarusian nuclear power plant, which could become a new potential source of nuclear and radiation hazards.

## 2. Sample data extraction

The development of a scientific justification for measures to ensure the radiation protection of irradiated individuals required the introduction of a new approach to the problem, and should now be based on the study of the types of exposure situations. According to the International Commission on Radiation Protection (ICRP) Publication 103, three types of situations are considered: planned exposure, existing exposure and emergency exposure [1]. For the Republic of Belarus, such an approach allows solving the tasks of ensuring radiation safety, including using innovative technologies and informatization in the conduct of radiation monitoring of the environment.

IOP Conf. Series: Materials Science and Engineering **487** (2019) 012001 doi:10.1088/1757-899X/487/1/012001

The main goals and objectives of studying and assessing the radiation situation using radiation monitoring can be viewed from two perspectives. The most common "pragmatic" approach is to assess the radiation situation for studying pathways of forming radiation doses, by including radionuclides in the trophic chains at different stages of their migration in the environment, i.e. to solve the pragmatic task of radiation monitoring [2].

Another global but less applicable approach is the study of the behavior of radionuclides as participants in the geochemical transfer of elements in the biosphere. This view of the problem allows solving the problems of ensuring radiation safety and protection not only of the population, but also of vulnerable environmental objects, i.e. provides a comprehensive biosphere approach.

The choice of the goal of studying and assessing the radiation situation: the solution of the pragmatic task of ensuring the radiation safety of the population or the implementation of an integrated biosphere approach is fundamental in terms of defining a strategy for organizing and managing a radiation monitoring system.

Currently, the use of an integrated biosphere approach in the Republic of Belarus is becoming a priority, because the state system for preventing and eliminating emergencies is designed to ensure effective protection of the population and territories from natural and man-made emergency situations, which should be adequately reflected in programs for radiation monitoring of the environment.

In addition, this approach is consistent with ICRP Publication 103, where, in addition to "practice and intervention" in the field of radiation protection, it is proposed to establish principles for proving sufficient environmental protection, which generally corresponds to the principles of an integrated biosphere approach [1].

Scientific studies of the consequences of nuclear weapons tests on the environment have created a new level of scientific understanding of the behavior of radioactive elements of artificial origin in the biosphere, in comparison with classical geochemistry, and radioactive contamination as a result of the Chernobyl disaster required their further development.

In the works of Yu A Izrael, F I Pavlotskay, K P Makhanko, C I Bobovnikova, S M Vakulovsky, V A Vetrov, V M Prokhorov, E D Shagalova, A N Marey, R M Alexakhin and others performed before 1986, the main attention was paid, first of all, to the study of radioactive fallouts from the atmosphere and their spatial distribution throughout the territory of the Soviet Union as a whole and the Republic of Belarus in particular, the so-called. global bomb contamination.

In rhe results of their research have been created a scientific basis for solving the "pragmatic" task of estimating radiation doses. On the other hand, based on classical geochemistry, they obtained results that allowed creating a scientific basis for describing the behavior of such radionuclides as cesium-137 and strontium-90 in the environment.

The results of studies on the migration of cesium-137 and strontium-90 global origin due to nuclear weapons tests in soils were successfully applied in the study and minimization of radioecological consequences of the Chernobyl catastrophe.

In the Republic of Belarus, studies of global bomb contamination conducted prior to the Chernobyl accident, including the work of E D Shagalova, made it possible to assess the radiation situation on the territory of Belarus for this period. The following parameters were studied: the exposure dose of gamma radiation, soil contamination with cesium-137 and strontium-90, the characteristics of their vertical migration in the soil.

It should be noted that this approach corresponds to the established world practice of ensuring radiation safety and is reflected in the IAEA Glossary under the definition of "natural analogue". According to [2], "the situation arising in natural (natural) conditions, which is used as a model for processes affecting technogenic systems, and allows you to draw the conclusions necessary to draw conclusions about the safety of an existing or planned nuclear installation. In particular, these can be mineral deposits containing radionuclides whose migration history, covering very long periods of time, can be analyzed, as well as the results used in modeling the potential behavior of these or similar radionuclides in the geosphere for an extended period of time.

After the Chernobyl catastrophe, a wide circle of specialists from the most affected countries, the Republic of Belarus, Ukraine, the Russian Federation, has been and continues to be engaged in studying and assessing the radiation situation, analyzing regional features of radionuclide behavior in the environment and investigating the mechanisms of their migration in the biosphere.

E P Petryaev, I M Bogdevich, V I Ternov, S K Firsakova, E D Shagalova, M G Germenchuk, O M Zhukova, V Yu Ageyets, G A Sokolik, A M Dvornik, Yu. M Zhuchenko et al created the foundation modern system of radiation monitoring of the environment in the Republic of Belarus.

Initially, studies of Chernobyl contamination were based on previously formed scientific views and approaches to solving the "pragmatic" task of assessing radiation doses, mainly from strontium-90 and cesium-137, but the scale of pollution and the severity of radioecological consequences required a new approach to assessing the radiation situation.

Consider the need to improve methodological approaches to assessing the radiation situation after the Chernobyl disaster, taking into account environmental pollution by iodine-131 [3].

It should be noted that during the period of nuclear weapons tests in the atmosphere, the study of the behavior of radioactive iodine-131 in the environment was not a top priority, due to a short half-life, it almost completely disintegrated in the troposphere and stratosphere, and its insignificant quantities that reached the surface, due to the global nature of the process, were almost completely absorbed by the world's oceans. Therefore, the task was not to assess the dose loads from radioactive iodine, including iodine-131, especially taking into account regional geochemical features of individual territories.

At the same time, research carried out in the 70s of the 20th century, L I Lozovsky on the study of regional features of the distribution of stable iodine in the soils of Belarus made it possible to estimate the content of iodine in one of the most important reservoirs of the biosphere-soil, to reveal the spatial variability of its concentrations, to identify the zones of so-called iodine deficiency, and, accordingly, the "iodine" deficiency in food and drinking water.

From the point of view of the theory of risk, these geochemical features of soils increased the vulnerability of the population of Belarus in the event of accidental pollution of the environment by radioactive iodine.

These geochemical and, accordingly, radiological assessments of risks could be used as a basis for the iodine prophylaxis of the population in the first period of the Chernobyl catastrophe during the so-called "iodine strike" and thereby significantly increase its effectiveness.

As is known, WHO recognized cases of thyroid cancer induced by radioactive isotopes of iodine due to the Chernobyl accident, the most serious consequences of the disaster for public health. In Belarus, after the Chernobyl disaster, over 1,500 cases of thyroid cancer in children and adolescents have been reported. Specialists note that the deficit of stable iodine in the environment and, accordingly, in foodstuffs creates the prerequisites for aggravation of the effects of radioactive isotopes of iodine on the thyroid gland, which means the presence of potential radiological risks.

The non- homogenous distribution of radioactive iodine throughout the territory of Belarus and, correspondingly, the incidence of thyroid cancer in children and adolescents, as well as the difficulties in calculating dose loads on the thyroid gland, required the use of a comprehensive biospheric approach, including the regional background of stable iodine. In addition, at the subsequent stages of assessing the radioecological consequences of the disaster, the lack of actual data on the pollution of environmental objects with radioactive iodine isotopes throughout the country required the use of calculation methods for the reconstruction of depositions and dose loads.

As already noted, radionuclides of artificial origin, entering the environment, are included in biogeochemical natural cycles, are involved in the geochemical transfer of elements between reservoirs of the biosphere and at the same time disrupt the structure of their natural elemental composition and balance.

The transfer of radioactive elements of iodine, primarily iodine-131 and iodine-129, which occurred as a result of radioactive contamination of the environment as a result of the Chernobyl disaster, is such a breach of balance.

In April-May 1986, the so-called "fresh" fission products of uranium, including iodine radionuclides, appeared in the atmosphere. These radionuclides, due to advection and atmospheric dispersion processes, moved in the atmosphere and then fell onto the underlying surface, including soil, vegetation, surface water, and entered artificial objects.

Thus, the natural balance of iodine content in individual reservoirs of the biosphere (surface layer of the atmosphere, surface water, biota, etc.) was disturbed, since its quantity increased due to radioactive, including short-lived isotopes.

In the general case, according to V I Vernadsky, each atom, having passed four forms of finding chemical elements (rocks, magma, biotic objects and scattered form), eventually returns to the beginning of the cycle. Moreover, the existence of such cyclic processes obeys the theory of "equilibrium of inhomogeneous media", proposed in the 19th century by V Gibbson, which relates the mechanism of transfer of elements to physical and chemical processes.

From a methodological point of view, it is important to understand that the life cycle of individual atoms can be infinite or finite. And it is the radioactive isotopes of chemical elements that have finite cycles, which is determined by their fundamental property of spontaneous radioactive decay and, accordingly, the half-life of each of the radionuclides.

That is why the amount of radioactive iodine in individual biosphere reservoirs depends on the radiation-chemical transformation of radionuclides in the reservoir, from the time, as well as the rate of their migration in the reservoirs and between them. The contribution of radioactive iodine to the planetary balance depends on the half-life of individual iodine radionuclides.

From this, one can make the conclusion that the time of radioactive isotopes stay in the biosphere and its individual reservoirs is extremely important for planning programs and evaluating the results of radiation monitoring. And the time of their stay in individual reservoirs of the biosphere and biotic objects, however, as well as in the biosphere as a whole, depends on their physico-chemical and radiation properties. The problems associated with the monitoring of short-lived radioactive isotopes of iodine in 1986, and, accordingly, the problems of estimating the doses of thyroid irradiation, are a vivid confirmation of this.

It should be noted that according to V I Vernadsky, a human is a "geochemical force" not only due to the passage of matter through his organism, but also due to the influence of anthropogenic activity in general on the flow and change of natural geochemical processes.

A vivid illustration of such influence on the environment can be the impact of agriculture, energy and transport, mining of minerals, primarily hydrocarbon energy. For the Republic of Belarus, potash mining, the chemical industry and related environmental problems, as well as an accident at the Chernobyl nuclear power plant, whose consequences for the country are assessed as an "ecological" catastrophe, are also very significant.

One of the significant sources of additional involvement of radionuclides in biogeochemical cycles is the widespread use of organic and mineral fertilizers and various chemical meliorants with an increased content of natural radionuclides, for example, the use of potassium fertilizers containing in their composition up to 0.01% radioactive potassium-40 in a mixture of stable isotopes potassium-39 and 41.

As it was mentioned above, the system of radiation monitoring of the environment should provide a solution to two tasks: to estimate the doses of population exposure and to assess the state and radioactive contamination of the biosphere.

The key to a common solution to these two problems is the use of an integrated biosphere approach. Its essence lies in the fact that the mechanism of radioactive contamination of the environment is described from the point of view of natural and anthropogenic processes occurring in the biosphere, which makes it possible to evaluate and predict the pollution of both the biosphere as a whole and its individual objects, including man.

Thus, the concept of radioactive contamination of the environment can be defined as "the involvement of radioactive, including artificially obtained, isotopes in the cyclic transfer of elements in the biosphere through natural geochemical and anthropogenic processes".

According to the IAEA Glossary, radioactive contamination is defined as "radioactive substances present on surfaces or inside solid materials, liquids or gases (including the human body) where their presence is not expected or desirable, or the process leading to their presence in such places" [2].

This definition allows us to study the mechanisms of radionuclide behavior in the environment and their geochemical transfer. As examples of such involvement in the geochemical transfer and creation of new sources of radioactive contamination, one can consider the accumulation of radioactive cesium-137 in the so-called geochemical barriers in the soil or accumulation of radionuclides in bottom sediments of surface water bodies. In either case, the specific activity of radionuclides increases in a separate biosphere reservoir in comparison with adjacent reservoirs, which creates the prerequisites for the emergence of a new potential or active source of pollution. It is on these premises that the description of the mechanisms of radioactive contamination of groundwater and groundwater after the Chernobyl catastrophe is based.

Another example of the emergence of new sources of radioactive contamination in the biosphere after the Chernobyl disaster is the process of the appearance and accumulation of americium-241 in separate biosphere reservoirs, which is the product of the decay of plutonium-241 and was initially absent from the accidental release. However, now its specific activity in environmental objects is increasing, including in those "places" where, according to [2], its presence was not expected or undesirable, thereby creating a new source of additional irradiation.

In the Republic of Belarus, a radiation monitoring system has been established and is functioning, which is part of the national environmental monitoring system of the Republic of Belarus (NSME). The NSME includes a wide network of observation points and accredited laboratories. The main objects of radiation monitoring are atmospheric air (ground layer), soils, including agricultural and forest, surface and groundwater, animal and plant objects.

## 3. Results

In the first years after the catastrophe at the Chernobyl NPP, a survey of almost the entire territory of the Republic of Belarus was carried out. Radiation monitoring was carried out in all settlements of the Republic of Belarus (more than 20 thousand), on all agricultural and forest lands. Soil samples were taken out and analyzed to assess the contamination of cesium-137 (figure 1) and strontium-90. In the southern part of the country, in the 100-km zone of the Chernobyl NPP, the radioactivity of plutonium isotopes was measured, and subsequently, radioactivity of americium-241. When creating maps of radioactive contamination and predicting its changes, modern information technologies are used, for example, various GIS, geo-oriented databases, etc. [4].

For the period 1986 - 2016 years more than 635,4 thousand soil samples were taken, more than 1 million 519 thousand measurements of the dose rate of gamma radiation at the sampling sites and determinations of the specific activity of individual isotopes were carried out.

Based on the survey of the territory of the country in the first period after the disaster in 1986 - several times, then - annually, subsequently - once in five years, maps of radioactive contamination of the territory of Belarus with cesium-137, strontium-90 and isotopes of plutonium, as well as prognosis maps of cesium-137 contamination.

Based on these data, a list of settlements "Forecast of changes in the radiation situation in populated areas located in radioactive contamination zones" has been prepared, where a forecast of radioactive contamination of soils in individual settlements by cesium-137, strontium-90 and plutonium isotopes is given.

Today one of the most important tasks of ensuring radiation safety in the Republic of Belarus is the introduction of innovative technologies in conducting observations of the radiation situation at the Chernobyl and Belorussian nuclear power plants. For this purpose, an automated system of radiation monitoring has been created and successfully functioning in Belarus, which includes measurements of dose rate of gamma radiation in air, radioactive iodine-131 in the near-surface atmosphere on-line, monitoring of radionuclide content in environmental objects: atmospheric air , surface water, soil, as well as observations of chemical pollution of atmospheric air, surface water and hydrometeorological parameters [5].

A mandatory requirement is the provision of quality control in measuring laboratories in accordance with ISO / IEC 17025-2001.

IOP Conf. Series: Materials Science and Engineering **487** (2019) 012001 doi:10.1088/1757-899X/487/1/012001

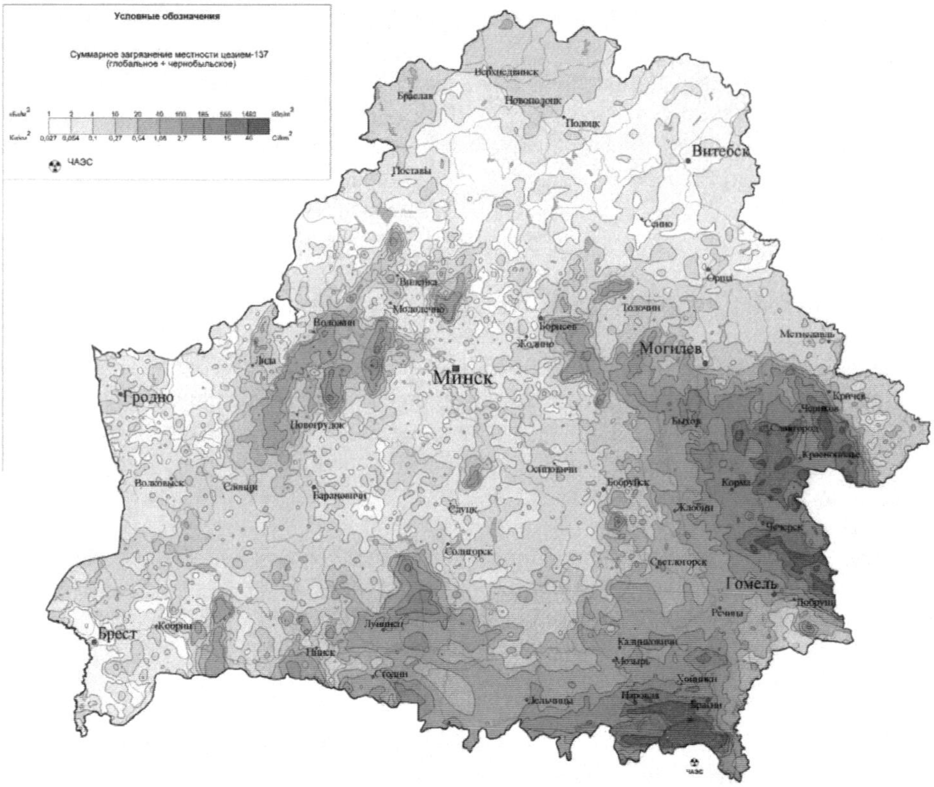


**Figure 1.** Radioactive contamination of soil by cesium-137 in the Republic of Belarus (May 1986).

This approach makes it possible to assess the radiation situation in on-line mode, as well as to carry out the forecast of radioactive contamination of atmospheric air, surface waters and soils taking into account real currently hydrometeorological conditions. For such calculations, certified computer codes are used.

## 4. Conclusion

Currently, the Republic of Belarus has a modern system of radiation monitoring of the environment, which is based on innovative technologies, including information technologies. After the accident at the Chernobyl NPP, the Republic of Belarus has the good practices in organizing radiation monitoring of the environment. With the use of GIS technologies maps of radioactive contamination of the territory of Belarus with cesium-137, strontium-90, isotopes of plutonium and americium-241 have been created.

New methodical approaches to the formation of a radiation monitoring system were developed in the event of new threats to radiation safety, for example, the construction of nuclear power plants. Based on the experience of studying the behavior of radionuclides, integrated biospheric approach to the organization of radiation monitoring of the environment is proposed.

Today we can say that the Republic of Belarus has a modern radiation monitoring system that is based on modern innovative technologies and allows ensuring the radiation safety of the population and the environment at a socially acceptable level.

## References

[1] *ICRP 2007 The 2007 Recommendations of the International Commission on Radiological Protection. Publication 103* (Amsterdam: Elsevier)

[2] *IAEA 2007 IAEA Safety glossary: terminology used in nuclear safety and radiation protection: 2007 edition* (Vienna: International Atomic Energy Agency)

[3] Straume L, Anspaugh R, Marchetti A A, Voigt G, Minenko V, Gu F, Men P, Trofimik S, Tretyakevich S, Drozdovitch V, Shagalova E, Zhukova O, Germenchuk M and Berlovich S 2006 Measurement of $^{129}I$ and $^{137}Cs$ in soils from Belarus and reconstruction of $^{131}I$ deposition from the Chernobyl accident *Health Physics* **91** 7–19

[4] Germenchuk M *et al* 1998 *Atlas of Caesium Deposition on Europe after the Chernobyl Accident* (Luxemburg: Publication of the European Communities) 65 p

[5] *IAEA 1997 Method for the Development of Emergency Response Preparedness for Nuclear or RadiologicalAaccidents. IAEA-TECDOC-953* (Vienna: International Atomic Energy Agency)

# Some results of determining the source and reasons for the appearance of 106Ru in Russia in September-October 2017

**V Shershakov and R Borodin**

RPA “Typhoon” of Roshydromet, Obninsk, Russia

**Abstract.** Probable locations of the sources of ruthenium-106 in the atmosphere in September-October 2017 have been considered. Taking into account the monoisotopic nature of the increase in ruthenium-106 concentration in the ground level air and atmospheric precipitation, the most probable version is substantiated, which suggests that ruthenium-106 came from a ground source located in the Southern or the Middle Urals. In order to locate the possible point of Ru-106 emission into the atmosphere, calculations of the time-inversed travel of air masses at the heights of 500-1500 m from the Ru-106 concentration measurement locations at European points during the relevant observation period were used. To confirm the correctness of the source location, calculations were performed that involved simulating regional transport and propagation of radioactivity in the near zone in case of a hypothetical emission of ruthenium-106 from a conventional source in the Southern Urals, in order to compare the simulation results with the available daily ruthenium-106 drop-out measurement data.

## 1. Introduction

According to an IAEA report [1, 2], in the period from September 25 to October 9, 2017, a radionuclide of ruthenium-106 in concentrations from 10 mBq/m$^3$ to 100 mBq/m$^3$ was detected in air samples of aerosols in the atmospheric air in the territory of European countries. The maximum concentration of 145 mBq/m$^3$ was registered in Romania. Such concentrations of ruthenium-106 in the atmospheric air do not pose a radiation hazard to the population and the environment. Nevertheless, the established international practice in the field of radiation monitoring related to such situations requires clarification of sources and reasons for the increase in the content of radionuclides in the environment. It should also be mentioned that the concentration of ruthenium-106 in air samples largely depends on the sampling time (exposure time). In this case, the exposure time at different observation points varied widely from several hours to several days and even weeks, making it difficult to perform a quantitative comparison of the available observational data. The prolonged nature of sample exposure makes it impossible to determine the exact maximum concentration of ruthenium-106 in the ground level air aerosols and the period of time during which it was observed.

There are various situations in which radionuclides may enter the environment. This may be an abnormal situation at the radiation-dangerous object associated with the loss of control over the release of volatile radionuclide compounds, or accidental disposal of a radioactive source, for example, when remelting metal scrap. However, with regard to the case under consideration, not a single enterprise, neither in Europe nor in Russia, declared an abnormal emission of ruthenium-106 during its production.

In the absence of information about the source, German and French specialists calculated so-called reverse trajectories from the points where ruthenium-106 was registered in Europe and estimated the location of the possible source. Based on the data obtained, it was stated that the emission point of ruthenium-106 is located in Russia, in the Urals.

In this situation, Roshydromet which coordinates activities related to radiation monitoring carried out work to assess the situation of the appearance of ruthenium-106 in the ground layers of the atmosphere and its distribution across Russia and Europe.

## 2. Analysis of data obtained from monitoring the content of ruthenium-106 in the ground level air aerosols and atmospheric drop-out in the territory of Russia

The main sources of information on the radiation situation in the territory of the Russian Federation are the observational systems of Roshydromet and Rosatom State Corporation. The most informative type of observation is the observation of radioactive air aerosols by means of air-filtering units (AFU). Roshydromet AFUs are located around nearly all first category radiation-dangerous objects in the territory of Russia. Aerosols are sampled with an exposure of 5 days, mainly. Radioactive drop-outs are sampled using horizontal plates with a daily exposure.

Starting from September 23, 2017, an increase in the total activity of beta-emitting radionuclides ($\Sigma\beta$) was observed in the territory of the Chelyabinsk region in the atmospheric aerosols samples, and spectrometric analysis of these samples revealed increased content of ruthenium-106. At the end of September and during the first ten days of October, the presence of ruthenium-106 in the ground level air and atmospheric drop-put was recorded at observation points in the territory of the Chelyabinsk region, Bashkiria, Tatarstan, Ulyanovsk, Samara, Saratov, Volgograd, Leningrad, Murmansk, Kursk, Voronezh and Rostov regions and near Krasnoyarsk city.

## 3. Analysis of possible ground sources of ruthenium-106 in Russia and in European countries

High levels of ruthenium-106 in the ground level atmosphere were observed across large areas, and therefore the source can be located both in the territory where the maximum levels were observed and at a considerable distance from these territories, due to the result of regional or transboundary transition.

To estimate location of the possible source, calculations were made, which consisted in simulating time-reversed transition of tracers from virtual sources located at the measurement sites. The model tracks the trajectories of time-reversed motion of the tracer particles emitted during the active period of the virtual source. The concentration of tracer particles at a given time interval reflects the probability of the location of the source of contamination acting during this interval.

The intensity of virtual sources corresponds to the measurement results at each point, i.e. the total emission of tracers is proportional to the concentration integrated over time. The time when measurements start and end at each point determines the start and end of the active period of the corresponding virtual source, the duration of each virtual source is equal to the samples averaging time at this point. Tracer properties (sedimentation rate, deposition rate, decay constant and leaching coefficient) are equal to 0.

The calculation is performed using the model of atmospheric diffusion with a predetermined time step $\Delta t$. At each step, the field of the average (for the $\Delta t$ interval) tracer concentration $C(x, y, h)$ is calculated.

The amount of tracer in a random elementary volume $\Delta V$ of the calculation area

$$q = \int_0^{\Delta V} C(x, y, h) \times dv$$

At the same time, the total number of tracer in the calculation area at this step

$$Q = \int_0^{V} C(x, y, h) \times dv$$

The value of $p = q/Q$ can be interpreted as conditional probability that the emission point is located in a given elementary cell at a given time interval, provided that the source is in the calculation area.

The calculation results may be conveniently presented in the form of a field of values of source location probability density.

$$f(x, y) = \frac{1}{Q} \int_0^{H} C(x, y, h) \times dh$$

where $H$ is the upper limit of the calculation area.

Then, by integrating the values of the $f(x, y)$ field over the area of a randomly specified area, we can calculate the conditional probability of source location in the specified area.

IOP Conf. Series: Materials Science and Engineering **487** (2019) 012002 doi:10.1088/1757-899X/487/1/012002

In this case, the *f(x, y)* integral over the area of the entire calculation region

$$\int_{0}^{S_0} f(x,y) \times ds = 1$$

i.e. the source is located in the calculation area.

Figure 1 shows an example of calculation results concerning the location of the source of ruthenium-106 as of September 25, 2017. The dots indicate the measurement points (virtual sources) providing the data according to which the calculation was carried out. The most significant measurement data were used with an exposure of not more than 2 days from 20 locations in Europe (according to IAEA data) and Roshydromet data (Argayash, Novogorny, Volgograd, Tsimlyansk) with an exposure of 5 days.

Using the calculation results it is possible to outline areas of the Southern and Middle Urals as a probable location of the source of Ru-106 release into the atmosphere.

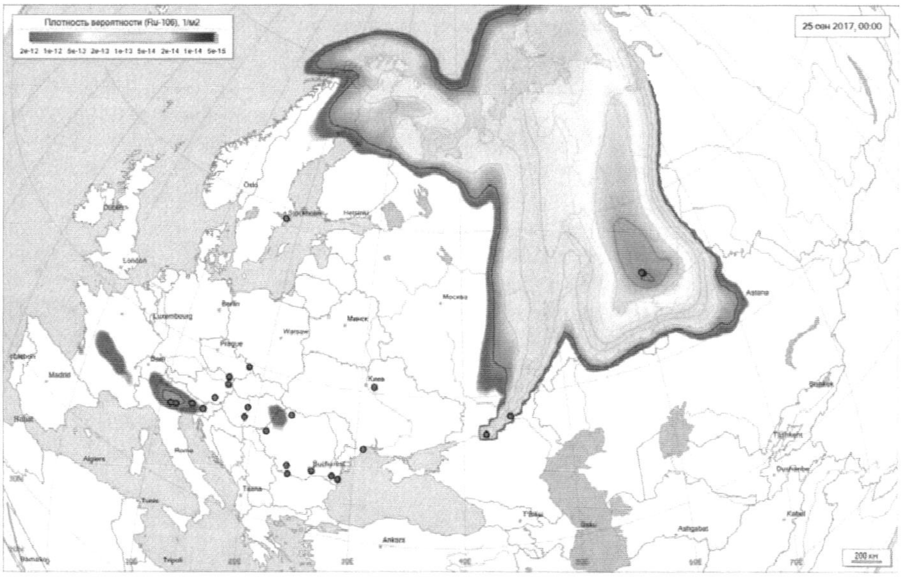


**Figure 1** – Probability density of ruthenium-106 source location on September 25, 2017, based on measurements in 20 points in Europe and in 4 points of Roshydromet.

## 4. Modelling dissemination of ruthenium-106 in the atmosphere from a hypothetical ground source located in the Urals

Scenarios were prepared and a series of calculations was performed to simulate dissemination of radioactivity in case of a hypothetical emission of Ru-106 from conventional sources in the Southern and Middle Urals in order to find the best match between the modelling results with the available data on Ru-106 concentrations at points in Russia and Europe.

The ruthenium-106 emission parameters varied in a wide range, namely: it began to release on September 23-24, 2017, the duration of the emission was from 2 to 7 days, the emission intensity was from 40 TBq/day to 250 TBq/day, the height of the emission was from 100 to 500 m.

Analysing the trajectories of the movement of air masses, it is possible to determine the time interval of the hypothetical ground source of ruthenium emission in the Urals region which is the closest match with the existing measurement data.

Figure 2 shows the calculation results of the ground level concentration of Ru-106 integrated for 10 days from a conventional source located in the Chelyabinsk region near the town of Ozersk.

Table 1 shows the results of comparison of the measurement data obtained at some points in European countries with the calculated data obtained from simulating a hypothetical emission from a conventional source located in the Chelyabinsk region near the town of Ozersk.

IOP Conf. Series: Materials Science and Engineering **487** (2019) 012002 doi:10.1088/1757-899X/487/1/012002

**Table 1.** Comparison of calculation results and measurement data related to ruthenium-106 concentration in the air.

| Country | Point | Start | End | Measurements (мБк/м$^3$) | Calculation, (мБк/м$^3$) |
|---|---|---|---|---|---|
| Austria | Graz | 02.10, 8:25 | 03.10, 10:16 | 36 | 30 |
| Austria | Graz | 03.10, 0:20 | 04.10, 7:50 | 13 | 10 |
| Austria | Graz | 04.10, 8:00 | 05.10, 7:00 | 1,0 | 1 |
| Czech Republic | Ostrava | 02.10, 6:00 | 03.10, 15:00 | 40 | 40 |
| Czech Republic | Ostrava | 03.10, 5:00 | 04.10, 15:00 | 1,1 | 1 |
| Hungary | Budapest | 25.09, 0:00 | 26.09, 0:00 | 2,7 | 0 |
| Hungary | Budapest | 26.09, 0:00 | 27.09, 0:00 | 1,7 | 0 |
| Hungary | Budapest | 27.09, 0:00 | 28.09, 0:00 | 2,0 | 0 |
| Hungary | Budapest | 28.09, 0:00 | 29.09, 0:00 | 1,9 | 0 |
| Hungary | Budapest | 02.10, 0:00 | 03.10, 0:00 | 27,6 | 40 |
| Hungary | Budapest | 03.10, 0:00 | 04.10, 0:00 | 9,9 | 30 |
| Hungary | Budapest | 04.10, 0:00 | 05.10, 0:00 | 2,1 | 3 |
| Italy | Udine | 02.10, 0:00 | 03.10, 0:00 | 49,1 | 10 |
| Italy | Udine | 03.10, 0:00 | 03.10, 0:00 | 54,3 | 15 |
| Italy | Udine | 03.10, 0:00 | 04.10, 0:00 | 30 | 3 |
| Italy | Udine | 04.10, 0:00 | 05.10, 0:00 | 5,2 | 1 |
| Italy | Udine | 04.10, 0:00 | 04.10, 0:00 | 4,2 | 1 |
| Romania | Bucuresti | 29.09, 0:00 | 29.09, 0:00 | 37,9 | 40 |
| Romania | Bucuresti | 30.09, 0:00 | 30.09, 0:00 | 145 | 150 |
| Romania | Bucuresti | 01.10, 0:00 | 01.10, 0:00 | 18,1 | 20 |
| Sweden | Stockholm | 30.09, 8:42 | 01.10, 8:42 | 0 | 0 |
| Sweden | Stockholm | 01.10, 8:42 | 02.10, 8:42 | 17 | 8 |
| Sweden | Stockholm | 02.10, 8:42 | 03.10, 8:42 | 9,8 | 10 |
| Switzerland | Lugano | 02.10, 0:00 | 02.10, 23:59 | 0,1 | 0,1 |
| Switzerland | Lugano | 03.10, 0:00 | 03.10, 23:59 | 1,9 | 0,4 |
| Switzerland | Lugano | 04.10, 0:00 | 04.10, 23:59 | 0,3 | 1 |
| Switzerland | Lugano | 05.10, 0:00 | 05.10, 23:59 | 0,5 | 0 |

**Figure 2.** The ground level concentration of Ru-106 integrated over 10 days according to the results of simulating Ru-106 dissemination in the atmosphere.

## 5. Modelling dissemination of ruthenium-106 in the near zone from a hypothetical ground source located in the Southern Urals

Calculations were performed that involved simulating propagation of radioactivity in the near zone in case of a hypothetical emission of ruthenium-106 from a conventional source in the Southern Urals, in order to compare the simulation results with the available daily ruthenium-106 drop-out measurement data.

When calculating the following parameters of the conventional source were adopted:

- source coordinates: 60.804° E, 55.696° N;
- start of emission: September 23, 00:00 (TAE);
- duration: 7 days;
- initial height of the emission: uniform distribution in the 50-100 m interval;
- source intensity: 40 TBq/day;
- dry deposition rate: $1 \cdot 10^{-5}$ m/s;
- calculation period: 7 days.

Figure 3 shows the results of calculating the total drop-out of ruthenium-106 over 7 days under given conditions.

Table 2 compares results of the simulation according to the conditions of the above scenario with the daily ruthenium-106 drop-out measurement data.

**Table 2.** Comparison of calculation results and measurement data related to daily drop-out of ruthenium106 obtained by Roshydromet RMN.

| Point | Measurements[a] (Bq/m$^2$) | Calculation (Bq/m$^2$) |
|---|---|---|
| Kyshtym | 51 | 50 |
| Argayash | 25 | 24 |
| Novogorny | 66 | 64 |
| Khudaiberdinsky | 32 | 41 |
| Metlino | 37 | 110 |

[a] For the period of 7 days exposure (September 23-30, 2017).

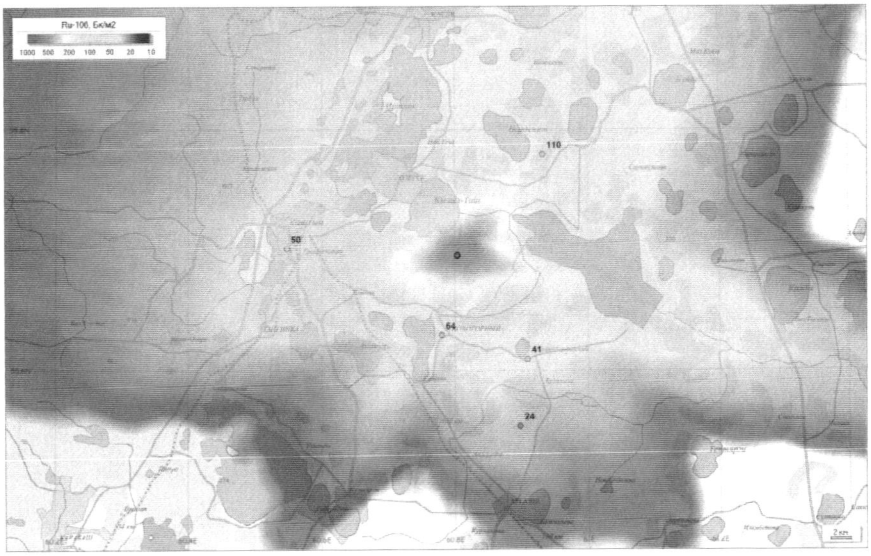


**Figure 3.** Integral drop-out of ruthenium-106 in the near zone over 7 days based on simulation results (the dots indicate the locations and values of ground measurements).

## 6. Conclusion

The results of modelling ruthenium-106 dissemination from conventional sources located in the Southern Urals (both on regional and local scale), taking into account the adopted values of the calculation parameters, show good agreement with the measurement data submitted by the IAEA and Roshydromet.

According to ruthenium-106 daily drop-out measurements, conducted by Roshydromet RMN (Radiation Monitoring Network), increased drop-outs were registered from September 23 to September 30, that is, the source remained active during this period. At the same time, the increased concentrations of ruthenium-106 in Europe (in the course of regional transition) are apparently due to the emission during the period from September 25 to September 26.

## References

[1] IAEA – International Atomic Energy Agency. Status of Measurements of Ru-106 in Europe. IAEA, Vienna, 2017. – 5 p.

[2] IAEA – International Atomic Energy Agency. Updated Technical Attachment Status of Measurements of Ru-106 in Europe. IAEA, Vienna, 2017. – 19 p.

IOP Conf. Series: Materials Science and Engineering **487** (2019) 012003 doi:10.1088/1757-899X/487/1/012003

# Calculation of critical parameters of the experimental stand «FKBN»

**V K Aznabaev, K D Kovalev, N A Mishchukov, V A Mishin and V V Kolesov**

Department of Nuclear Physics and Technology, INPI NRNU MIPhI, 1 Studgorodok, Obninsk, 249039, Russia

nickolai.mishukov@yandex.ru

**Abstract.** The experimental stand «FKBN» (physical fast-neutron boiler) is a critical assembly in which the vertical lift mechanism is used. With the help of such stands, in the second half of the 20th century, experiments were conducted on the effect of various materials in various designs on the kinetic characteristics of a system with fissile material. Mostly, such works were conducted in Los Alamos National Laboratory (USA), Oak Ridge Laboratory (USA), VNIIEF (Russia), VNIITF (Russia). On such installations, there were often accidents with the emergence of a self-sustaining chain reaction. One of these occurred on April 5, 1968 in VNIITF in Chelyabinsk-70 (now Snezhinsk). In this paper, the critical state of the stand is calculated for different distances of the reflector from the core under the conditions of the presence and absence of solid polyethylene inside the core. The configurations in which $K_{eff}$=1,000 and $K_{eff}$=1,000+$\beta_{eff}$ are of interest are of interest. As a tool, the CSAS6 sequence of the SCALE-6.2.1 software package was used.

## 1. Introduction

In the past, the exclusion of the possibility of the emergence of SCR was considered to be the main and the only noteworthy problem of nuclear safety. However, experience has shown that it is impossible to achieve absolute reliability of measures to prevent SCR, since economic indicators of production cannot be ignored. On the sum of potential consequences, the SCR is among the serious accidents: external exposure of personnel, release of radioactive substances and the possibility of external and internal exposure to a wider range of people, environmental pollution, equipment failure, deterioration and loss of valuable materials. However, the probability of an accident with such a sum of consequences is insignificant.

The need to obtain data on the course and consequences of SCR required special experiments. So, in the USA a number of critical assemblies were tested in conditions of instantaneous supercriticality (Godiva, Triga, KEWB, SPEPT, TREAT, etc.). However, the most significant was the French CRAC program, during which (1968–1972), more than 60 experiments were carried out with active zones from solutions of $^{235}$U with enrichment of 93%. Similar Soviet works also provided valuable information on the SCR kinetics and allowed the creation of a series of pulse reactors of self-extinguishing action.

It is now established that at the SCR is not a phenomenon that cannot be foreseen. Therefore, the analysis of the consequences must be done at the design stage of production processes and equipment. The most important measures are measures to counteract the radiation consequences of the SCR,

IOP Conf. Series: Materials Science and Engineering **487** (2019) 012003 doi:10.1088/1757-899X/487/1/012003

which, as a rule, are reduced to the design of an emergency alarm system, optimal evacuation routes, the use of protective equipment, emergency dosimetry and etc.

## 2. Accident with the emergence of SCR in Chelyabinsk-70, April 5, 1968

The accident occurred on April 5, 1968 in the Russian Federal Nuclear Center VNIITF (RFNC-VNIITF), located in the southern part of the Ural Mountains between Yekaterinburg and Chelyabinsk. Experiments with critical assemblies using the vertical lift mechanism at the FKBN facility began at VNIITF in 1957. The FKBN, in particular, implemented a number of critical configurations with a thick reflector and a large internal cavity providing operation in the pulsed mode and in a static mode at power several kW.

Stand BKBN consisted of a support, which housed the central hollow sphere, in fact is as. Above and below were hemispherical reflectors that were attached to the rods of mechanical drives. Changing with their help the distance and measuring the count of neutrons, it was possible to determine the degree of subcriticality of the system.

In the present case, the effect of the presence of a polyethylene spherical sample on the kinetic characteristics of the reactor system was investigated.

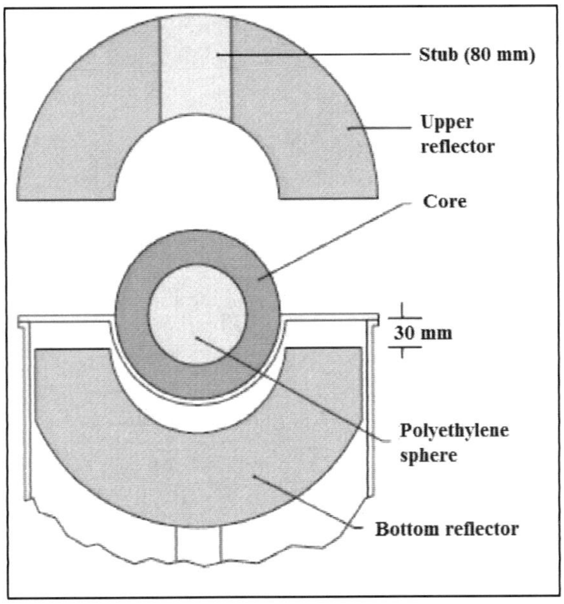


**Figure 1.** Schematic representation of the configuration of the vertical lift assembly and the FCSN core during the accident.

At the first stage of the experiments, inside the cavity of the core was empty, and critical assembly parameters were defined. In the second stage, a polyethylene sphere was placed inside the cavity and again the critical state was measured. To speed up the work (it was the 2nd evening shift on the eve of the weekend), it was decided to quickly lower the upper reflector to the core, and then use a fine calibration to raise the lower reflector until a critical condition arrived.

From the manual control panel, the responsible work supervisor began operating the upper hoist to lower the upper half of the reflector and bring it into contact with the core. The operator, standing next to the FKBN, guided a descending reflector with his hand. The accident occurred when the upper half of the reflector fell towards the core and almost came in contact with it. After the power of the ejection reached a kilowatt level, a high-speed emergency protection triggered and the lower half of the reflector was dropped, which was enough to bring the system into a deeply subcritical state and stop the chain reaction. The emergency release led to the death of two specialists standing next to the assembly during the release. The criticality alarm system was installed, but did not work during the experiment.

## 3. Description of the calculation model

Based on the description of the critical assembly, a calculation model was constructed. Its 3D image is shown in figure 2.

Parameters of the calculation model:

- Central hollow sphere with radii: 5.15 cm 9.15 cm
- Hollow hemispherical reflector (2 halves) with radii: 9.15 cm and 20.00 cm
- The central sphere consists of highly enriched uranium metal (90% of $^{235}U$)
- Reflector consists of metallic uranium with natural enrichment (0,71% of $^{235}U$)

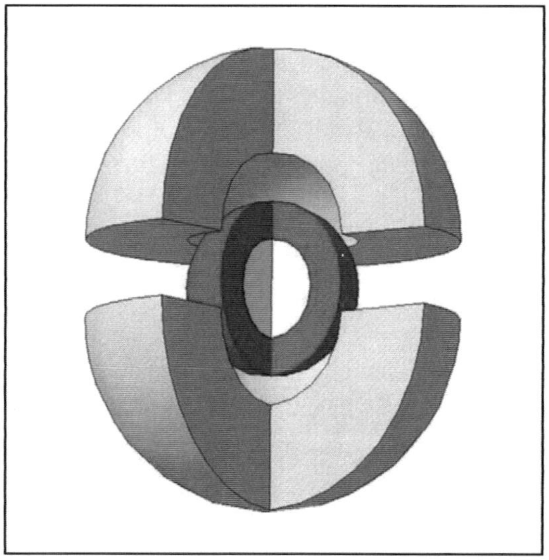

**Figure 2.** Critical stand model.

The calculation is performed using the CSAS6 module of the SCALE-6.2.1 software package, which uses the KENO-VI Monte Carlo code. ENDF / B-VII.1 is used as a nuclear data library.

## 4. Calculation

At the first stage, we determine the critical parameters of the system, namely the distance between the reflectors and the az. in the absence of materials in the cavity of the central sphere.

First we will change the distance between the core and the upper reflector, leaving the lower reflector in the initial position. The results of measuring the effective multiplication factor are shown in table 1.

**Table 1.** Effective reproduction rate and distance for the upper reflector.

| № | Distance R (cm) | Calculated value of $K_{eff}$ | Estimated accuracy ($\Delta K/K$) |
|---|---|---|---|
| 1.1 | 2,00 | 0,8875 | 0,0003 |
| 1.2 | 1,50 | 0,9007 | 0,0003 |
| 1.3 | 1,00 | 0,9153 | 0,0003 |
| 1.4 | 0,50 | 0,9311 | 0,0003 |
| 1.5 | 0,00 | 0,9484 | 0,0003 |

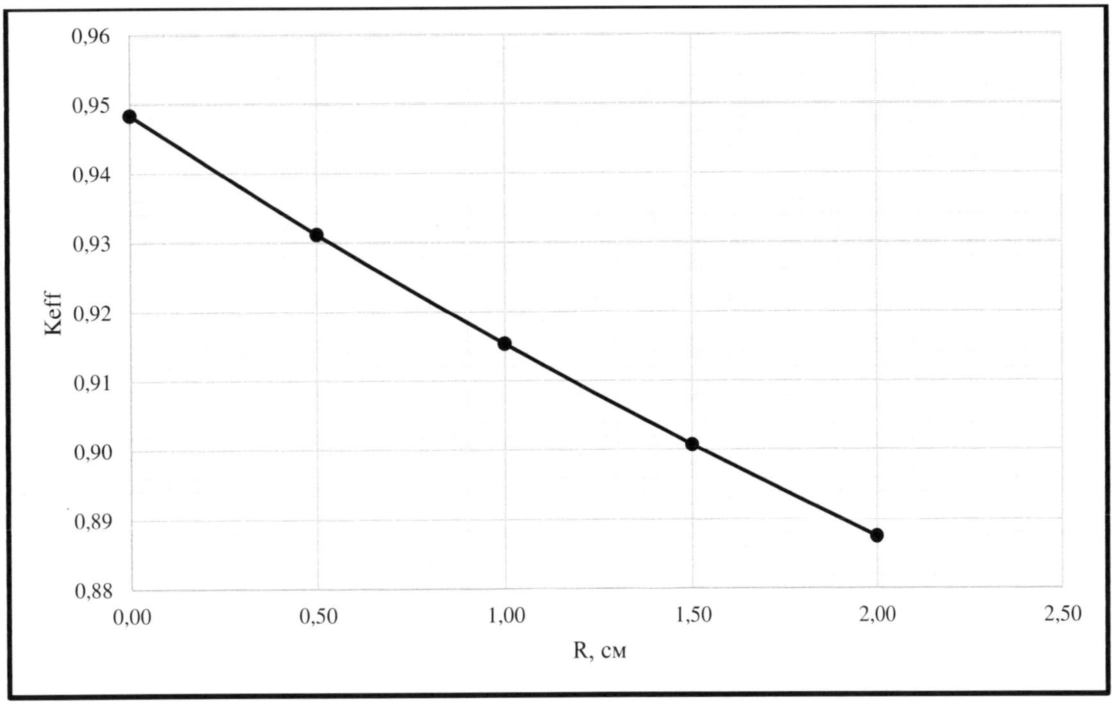


**Figure 3.** Dependence $K_{eff}$ (R) for the upper reflector.

Thus it can be seen that even a complete approach of the upper reflector and core does not lead to criticality. Therefore, in the following calculation, we will vary the distance between the core and the lower reflector, with the upper reflector completely lowered.

The results of the calculations are shown in table 2, and are also graphically shown in figure 4.

**Table 2.** Effective reproduction rate and distance for the lower reflector.

| № | Distance R (cm) | Calculated value of $K_{eff}$ | Estimated accuracy $(\Delta K/K)$ |
|---|---|---|---|
| 2.1 | 4,00 | 0,9280 | 0,0003 |
| 2.2 | 3,50 | 0,9381 | 0,0003 |
| 2.3 | 3,00 | 0,9484 | 0,0003 |
| 2.4 | 2,50 | 0,9597 | 0,0003 |
| 2.5 | 2,00 | 0,9725 | 0,0003 |
| 2.6 | 1,50 | 0,9854 | 0,0003 |
| 2.7 | 1,00 | 0,9996 | 0,0003 |

IOP Conf. Series: Materials Science and Engineering **487** (2019) 012003 doi:10.1088/1757-899X/487/1/012003

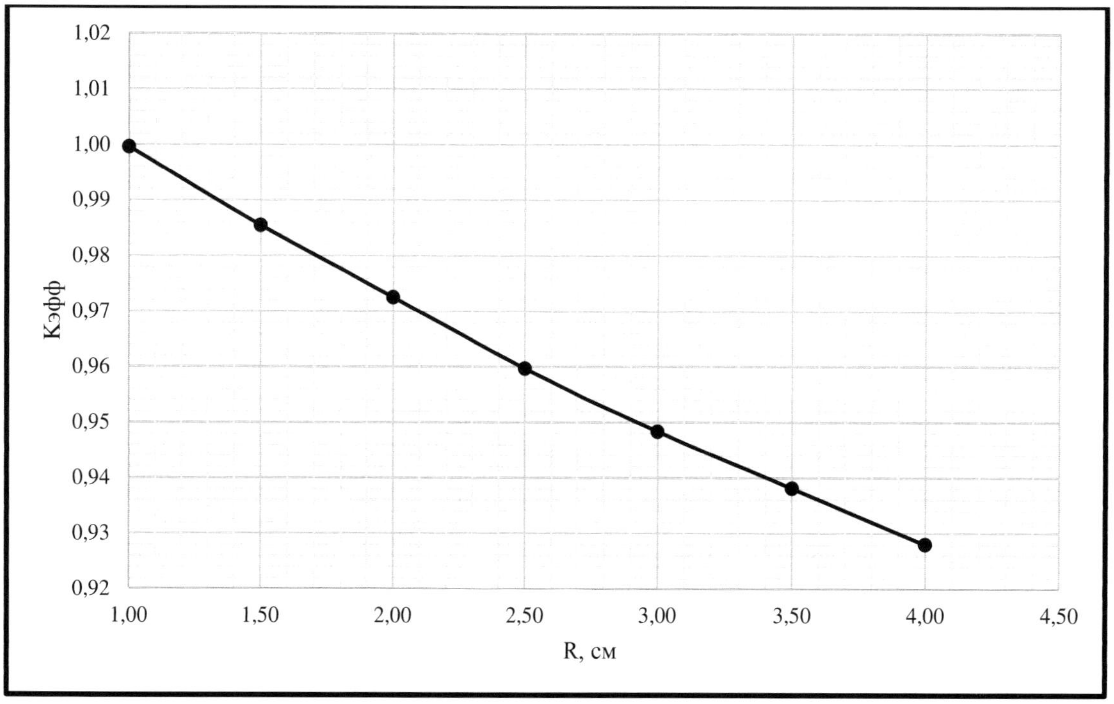


**Figure 4.** Dependence $K_{eff}$ (R) for the lower reflector.

From these results it is evident that in the absence of materials in the cavity of the core. Criticality will arise only when the upper reflector is completely lowered and when the lower reflector approaches a distance of 1 cm from the core

At the next stage of the calculations, we place inside the cavity of the core the solid polyethylene. Just as in the previous case, we will change the distance between the core and the upper reflector, leaving the lower reflector in the initial position.

The results of the calculations are shown in table 3, and are also graphically shown in figure 5.

**Table 3.** Effective reproduction rate and distance for the upper reflector.

| **№** | **Distance R (cm)** | **Calculated value of $K_{eff}$** | **Estimated accuracy ($\Delta K/K$)** |
|---|---|---|---|
| 3.1 | 2,00 | 0,9771 | 0,0003 |
| 3.2 | 1,60 | 0,9873 | 0,0003 |
| 3.3 | 1,20 | 0,9971 | 0,0003 |
| 3.4 | 0,80 | 1,0083 | 0,0003 |
| 3.5 | 0,40 | 1,0202 | 0,0003 |
| 3.6 | 0,00 | 1,0332 | 0,0003 |

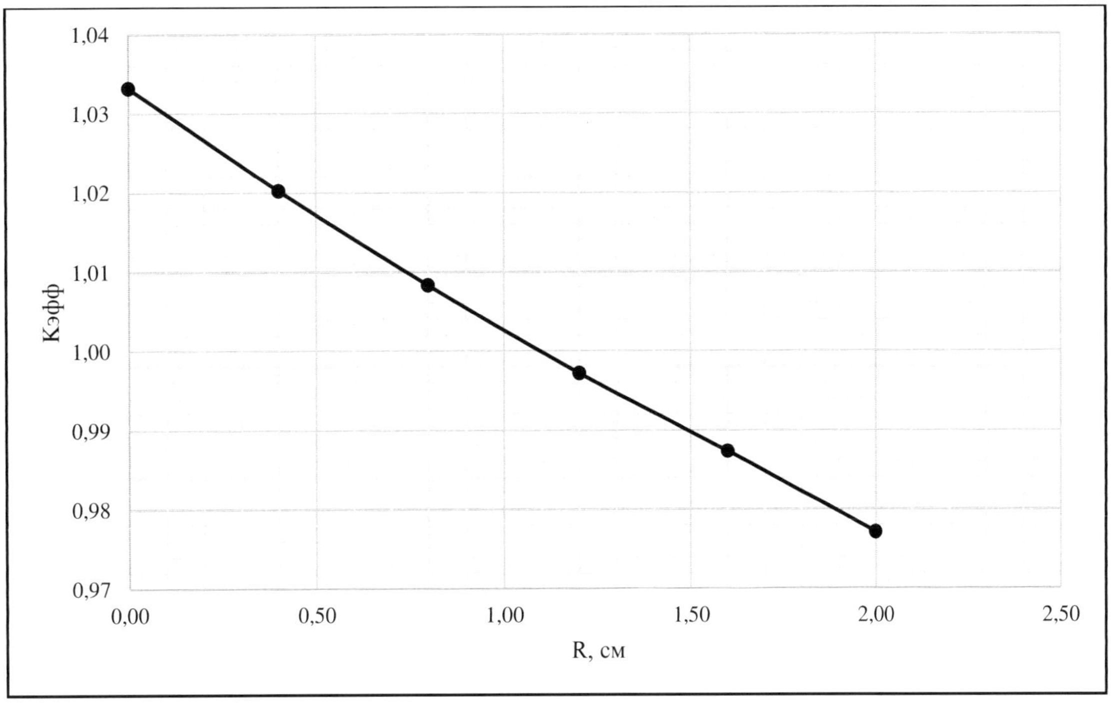


**Figure 5.** Dependence $K_{eff}$ (R) for the upper reflector.

It can be seen from the graph that even at a distance of 1.10 cm between the core and the upper reflector is critical. It is clear that further convergence of a core with the upper reflector will lead to a large supercriticality, and consequently to acceleration on the instant neutrons, even if the initial position of the lower reflector is preserved.

## 5. Conclusions

In the course of the work, the description of the accident with the emergence of the SCR occurred in VNIITF was studied. A computational model was constructed for calculating the critical parameters.

In analysing the results obtained, we can say that in the case of an empty core the criticality arises when the core with, the upper reflector and the lower reflector are completely approaching at a distance of 1.00 cm.

When installing polyethylene inside the core criticality occurs at a distance of 1.10 cm from the lower reflector to the core.

The initial placement of the lower reflector just below indicates the insurance of the experimenters. However, the blunder of the responsible manager is to underestimate the positive effect of polyethylene on the criticality of the stand. From the calculations, we obtain an additive of positive reactivity when adding polyethylene, which is equal to 10.33%.

## Reference

[1] Frolov V V, Macklafin T P, Monakhan S P, Pruvost N L and Ryazanov B G 2003 *Overview of Nuclear Accidents with the Emergence of SCR (LA-13638)* (Los-Alamos National Laboratory) p 236

[2] Petrie L M, Bekar K B, Hollenbach D F and Goluoglu S 2016. *CSAS6: Control Module for Enhanced Criticality Safety Analysis with KENO-VI* (Oak Ridge National Laboratory) p 217

# The emergence of technogenic risks in the event of a heat exchange crisis in nuclear power plants

**P A Belousov, D A Raspopov and K P Stepanchenko**

National Research Nuclear University of MEPhI (Moscow Engineering Physics Institute) Obninsk Institute of Atomic Energy (IATE), Studenchesky Gorodok 1, Obninsk, 249030, Russia

E-mail: raspopovdmitriyalexeevich@gmail.com

**Abstract**. In this article, the main problems and man-caused risks are considered in the event of a heat exchange crisis in nuclear power plants (NPP). A thermophysical booth simulating the operation of a nuclear reactor was developed. The data of acoustic diagnostics of the boiling of the coolant on the basis of the work of the laboratory stand were collected. Methods for processing and analyzing data collected from acoustic sensors have been developed. The recording and processing of signals was carried out using the application package and MATLAB programming language. The article describes the application of correlation and spectral analysis for data processing and analysis. The authors also use regression analysis to find the dependence of wall temperature on the frequency of acoustic oscillations. Forecasting the values of the wall temperature from the frequency of sound vibrations makes it possible in the future to find diagnostic features and build a mathematical model for detecting boiling and heat transfer crisis in nuclear power plants.

## 1. Introduction

Providing effective heat removal and avoiding overheating and burnout of heating surfaces, in particular fuel elements of nuclear reactors, are necessary conditions for safe operation and prevention of severe accidents with the release of radioactivity beyond the safety barriers at nuclear power plants.

A vivid example of an accident with a violation of the heat sink is an incident in 1979 at the Three Mile Island NPP, where the residual heat emission played a significant role. The cause of the accident was the average leakage from the first circuit, which led to a decrease in the pressure in the core and effervescence of the coolant on the residual heat release.

In emergency modes, when the pressure in the reactor decreases and the coolant flow through the active zone decreases, boiling of the coolant can begin. The boiling is the process of intense evaporation occurring in the entire volume of the liquid at the saturation temperature or somewhat superheated relative to the saturation temperature, with the formation of vapor bubbles.

At surface boiling, the temperature of the shell is higher than the boiling point of water (saturation temperature), and the temperature of the coolant is lower than the saturation temperature. In this case, a bubbling boiling mode is observed, where bubbles freely leave the surface of the shell and are carried away by the flow of water.

In this case, heat transfer only improves, as the heat transfer coefficient increases. However, if the heat flux locally increases markedly and the rate of bubble formation at this point increases, and the rate of their entrainment by the coolant decreases (for example, the flow through the zone decreases, or the deformation of the fuel grid), a steam film can form on the TVEL surface. With the formation of a local vapor film, the heat transfer coefficient falls by a factor of tens and the shell overheats, as a result of which a heat exchange crisis of the first kind occurs, as described in the book [1].

Thus, the first modification of the crisis (a crisis of the first kind) is interpreted as a consequence of the transition of the bubbling boiling of liquid into a film one. The reverse transition from film boiling to bubble is called a crisis of the second kind. Its peculiarity consists in the fact that the vapor film breaks down and boiling occurs throughout the volume of the liquid, the so-called bulk boiling. In the event of a heat exchange crisis on the surface of the TVEL envelope, both the first and the second kind, the shell begins to warm up to high temperatures, at which the chemical reaction of zirconium oxidation begins when interacting with water.

From all the above, it follows that the diagnosis of boiling and the heat transfer crisis, which is the boiling stage, refers to the classic problems of diagnostics at nuclear power plants. In his article [2], E.I. Nesis asserted that a theoretical analysis of numerous experimental data made it possible to detect the presence of a correlation between the thermal parameters and the characteristics of the sound accompaniment during boiling. This statement is the basis for requiring the development of passive acoustic diagnostics of heat transfer in boiling liquid.

Thus, the most important task for ensuring the safe operation of nuclear power plants is the creation of a comprehensive diagnostic system, with the possibility of automatically recognizing abnormal and pre-emergency thermal-hydraulic regimes in nuclear reactors.

## 2. Materials and methods

### *2.1. Thermophysical stand*

The developed thermophysical stand is shown in figure 1. The stand is a closed system under pressure consisting of several parts. A tank with purified water is the working fluid of the installation. Heating elements are installed in the tank. They simulate the operation of TVELs at nuclear power plants. During the operation of the heating elements, water boils, then the water is converted into steam and moves along the connecting pipes. Steam enters the turbine connected to the alternator. The turbine rotates under the steam pressure. Steam enters the refrigerator after the turbine. Steam condenses on pipes in which water circulates from the cooler vessel. The condensed steam returns to the tank [3].

The stand is also equipped with multiple sensors and instruments for measuring pressure, temperature and other characteristics. The stand is compact and mobile. This allows it to be easily transported and operated in any environment. This installation allows observing such phenomena as: different boiling regimes, evaporation processes, dependence of the boiling temperature of water on pressure, generation of electricity through a dynamo machine.

### *2.2. Data collection equipment*

Experimental data on diagnostics were obtained with the help of acoustic sensors and thermocouples installed in a laboratory bench. The data was then processed using the MATLAB software. Four microphones were used as acoustic sensors with a frequency range from 4 Hz to 20 kHz, a sensitivity of 0.25 mV / Pa, a dynamic range of 146 dB. The temperature range is from -400 to + 1200 ° C. The general view of the acoustic sensor is shown in figure 2.

Industrial thermocouples of the brand - TXA-108-4x11-0-KX-7 / 0.2-2000 were used as temperature sensors. Thermocouples are shown in figure 3. The range of measured temperatures is from 50 to 300 ° C, and the operating temperature range of the cable is from 5 to 200 ° C.

The ADC-LVDS module was used to collect data from acoustic sensors. The ADC module is shown in figure 4. The main characteristics of the ADC-LVDS are: the number of analog channels is 4, the range of the input voltage of the channel is ± 1V, the ADC is 12 bits in length, the conversion speed is up to 65 Ms / s, and the bandwidth is 315 MHz.

**Figure 1**. Thermophysical stand.

**Figure 2**. Acoustic sensor.

**Figure 3**. Thermocouple TXA-108-4h11-0-KX-7 / 0.2-2000.

**Figure 4**. ADC-LVDS.

The main task in diagnosing the heat exchange crisis is to diagnose and determine the boiling point. The main signals for diagnosing boiling are acoustic noises. They are taken from pressure waves and bubble bursts during boiling in nuclear power plants, as described in [4].

Correlation analysis was used to process and synchronize signals from acoustic sensors. Correlation analysis allows us to evaluate the temporal properties of a signal without using spectral analysis. For example, to estimate the rate of change or duration of a signal and temporarily correlate one signal with another. The mutual correlation function (VKF) determines the time coupling of two signals in time. The autocorrelation function (ACF) shows the statistical relationship of one random variable over time.

Spectral analysis consists of decomposing a signal into its frequency or spectral components and evaluating or measuring their spectral characteristics. The mathematical basis of the spectral analysis of signals is the Fourier transform. The allocation of the frequency of the regular components of a signal noisy with interference is the main purpose of the Fourier transform. A direct Fourier transform takes a description of the signal from the time domain to the frequency domain. The fastest Fourier transform (FFT) is most often used for spectral analysis. The signal can be decomposed into its component oscillations of different frequency and amplitude using a fast Fourier transform. The windowed Fourier transform is a function of time, frequency, and amplitude. Window Fourier transform allows to receive the characteristic of frequency distribution of a signal with amplitude in time.

## 3. Results and discussion

### *3.1. Correlation analysis*

The MATLAB application package was used for the experiment. The "recorder" function built into MATLAB was used to record a signal from an acoustic sensor. Each signal is given its own time interval, depending on its length. Temporal signal implementations are shown in figure 5. Correlation analysis was used to process and synchronize signals from acoustic sensors using the methods and algorithms described in the book [5].

Synchronization of signals was carried out as follows. If there are delays between the signals, the cross-correlation function will have a maximum in the non-zero value. The most lagging signal is searched for and synchronized with the rest and the delay values are removed. The values of the delays were found and removed using a mutual correlation function. These values are shown in figure 6.

Timed realizations of signals after their synchronization are shown in figure 7. Further analysis of time series is possible only after carrying out this procedure.

### *3.2. Spectral analysis*

Spectral analysis of signals was carried out after their synchronization using the methods and algorithms described in the book [6]. The signal spectrogram is used to estimate the signal change over time. To calculate the spectrogram, the signal vector is divided into segments. The spectrum is calculated for each segment using the "fft" function (Fast Fourier Transform (FFT)). The FFT decomposes the signal into its component oscillations of different frequency and amplitude. A set of spectra of all segments forms a spectrogram. The built-in function in MATLAB "spectrogram" is used to calculate the signal spectrogram.

The spectrogram function is used to plot the spectrograms of each synchronized signal. The size of the "window" (the number of samples per second) should be equal to the sampling frequency, i.e. 44100. The number of samples for a discrete Fourier transform was chosen taking into account the window size and the acceleration of the FFT process. This value must be 2n and not exceed the window sizes. It follows that the value 215 = 32768 is the most appropriate window size for the FFT.

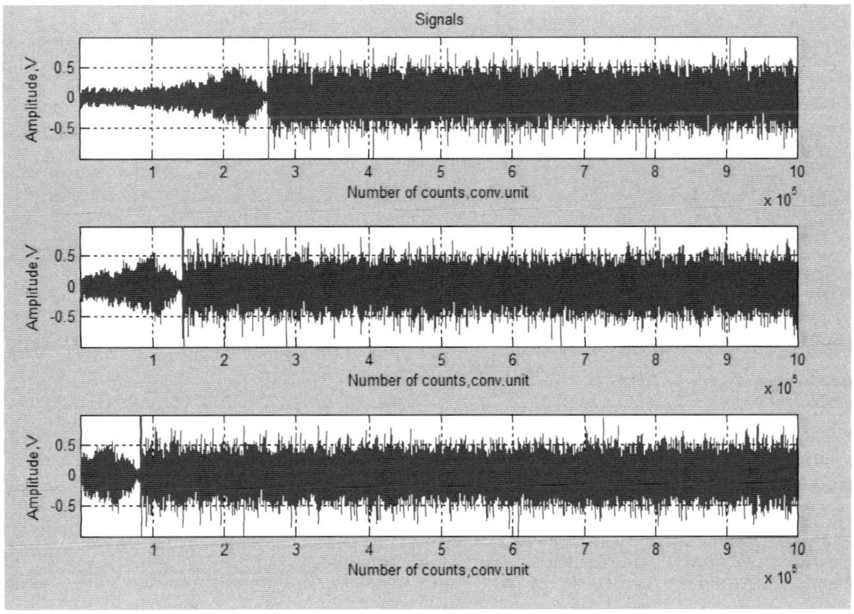


**Figure 5.** Time Signal Realizations.

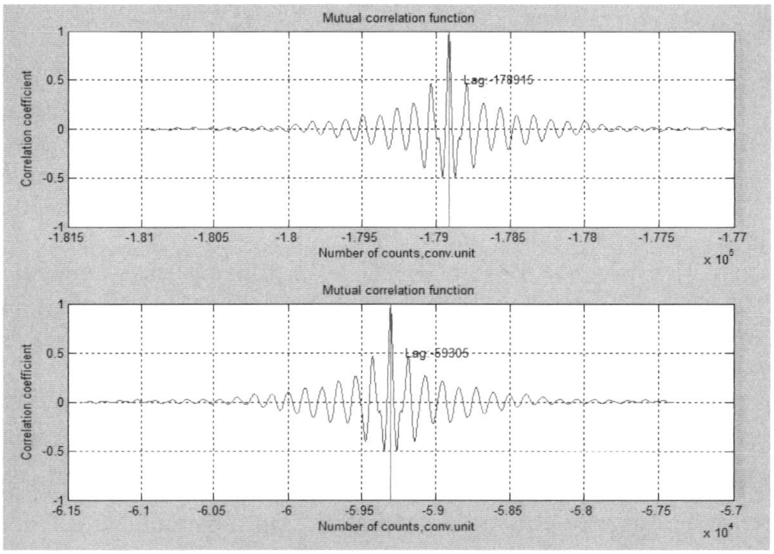


**Figure 6**. Cross correlation function of non-synchronized signals.

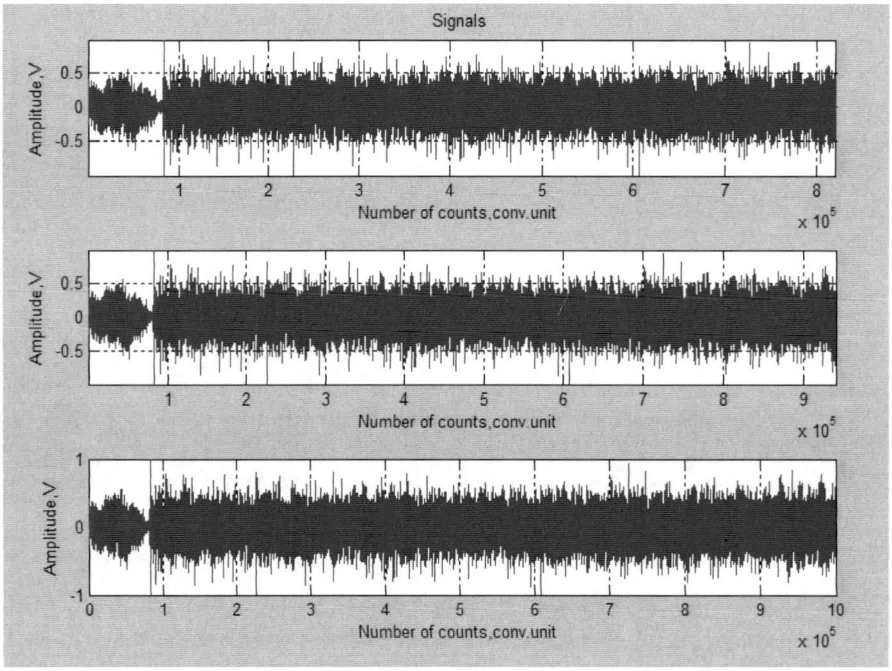


**Figure 7**. Timed implementations of synchronized signals.

Spectrograms of signals from acoustic sensors are shown in figure 8. By spectrographs, one can observe the boiling process, its duration, and also the frequency and amplitude of the acoustic oscillations of the bubble collapse.

For the construction of power spectra, MATLAB uses the Fast Fourier Transform algorithm. The power spectra for signals from acoustic sensors are shown in figure 9. The spectra show the distribution of signal power as a function of frequency.

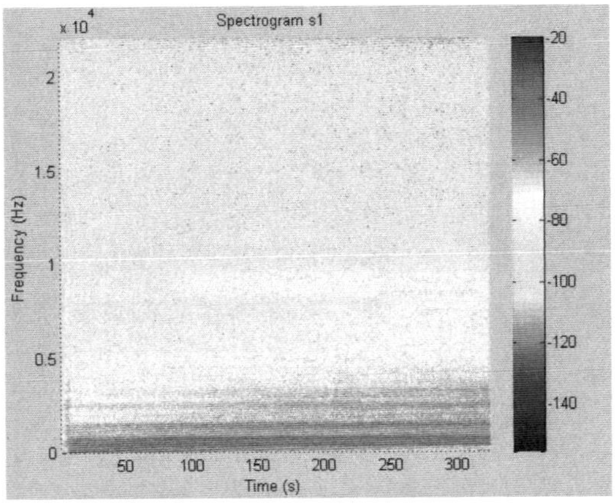


(a) s1

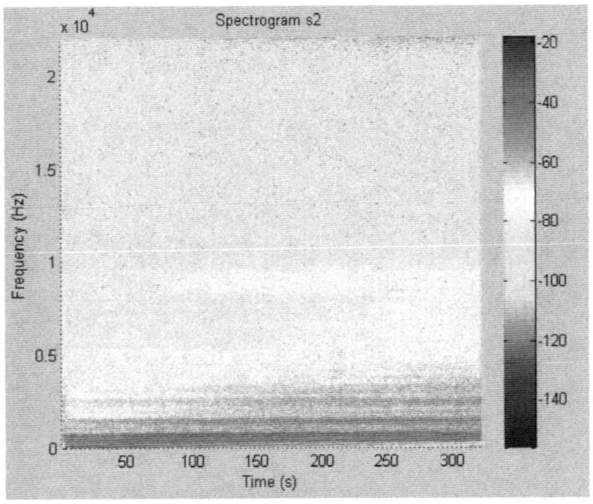


(b) s2

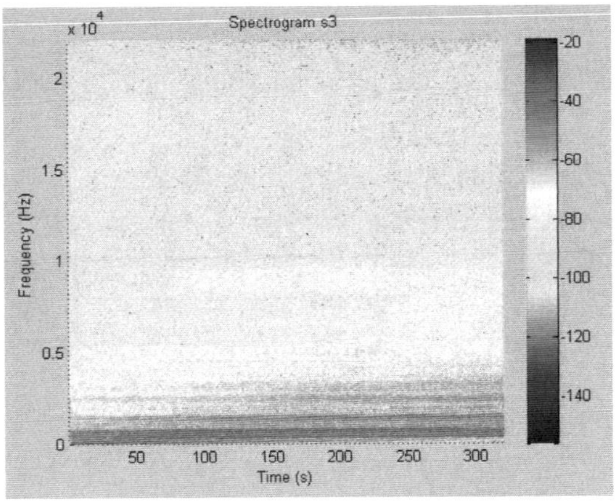


(c) s3

**Figure 8**. Spectrograms of signals from acoustic sensors; (a) s1, (b) s2, (c) s3.

The main task was to find the dependence of the wall temperature on the frequency of acoustic oscillations. The calculation of the correlation coefficient says k = 0.45, that such a relationship exists. Such a dependence was found by isolating the frequencies at which the signal power was maximal and the thermocouple measurements installed on the laboratory bench. Dependence on the frequency of sound vibrations and the wall temperature is linear with the help of OLS. The graph of the dependence of the wall temperature on the frequency of acoustic oscillations is shown in figure 10.

As can be seen from the graphs, the values of the maximum frequency powers will shift to low frequencies with increasing wall temperature, and the line will decrease with increasing temperature, as shown in figure 10. In the future, to predict the wall temperature from the frequency of sound vibrations at normal and other pressure it is necessary to use more complex regression models.

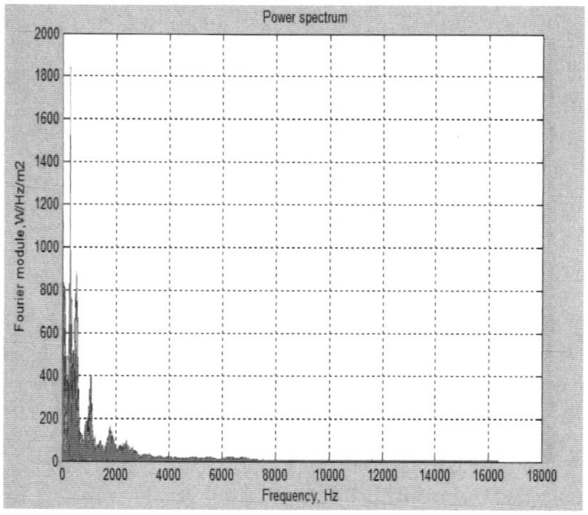


(a) s1

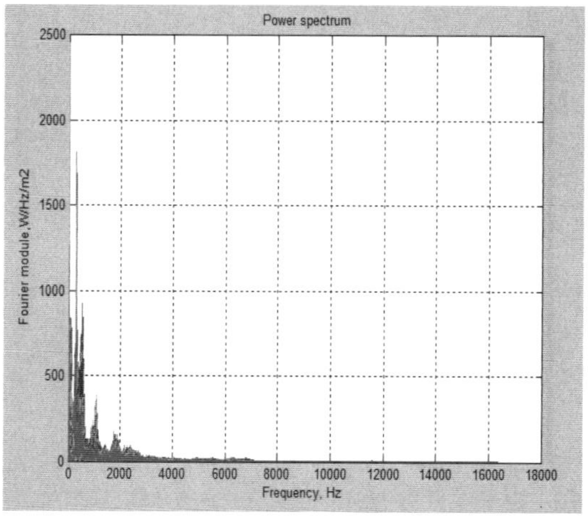


(b) s2

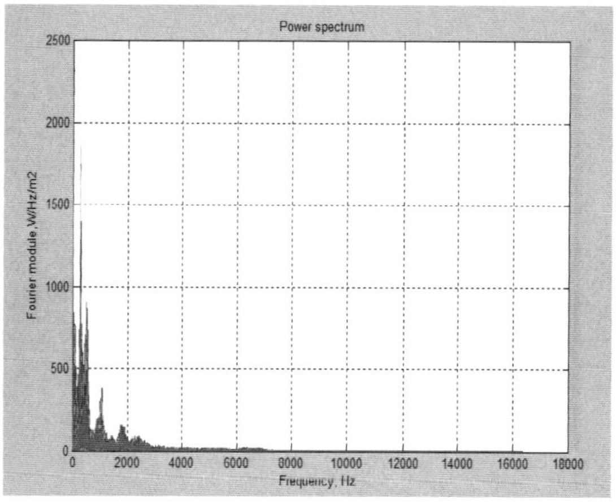


(c) s3

**Figure 9**. Power spectra for signals from acoustic sensors; (a) s1, (b) s2, (c) s3.

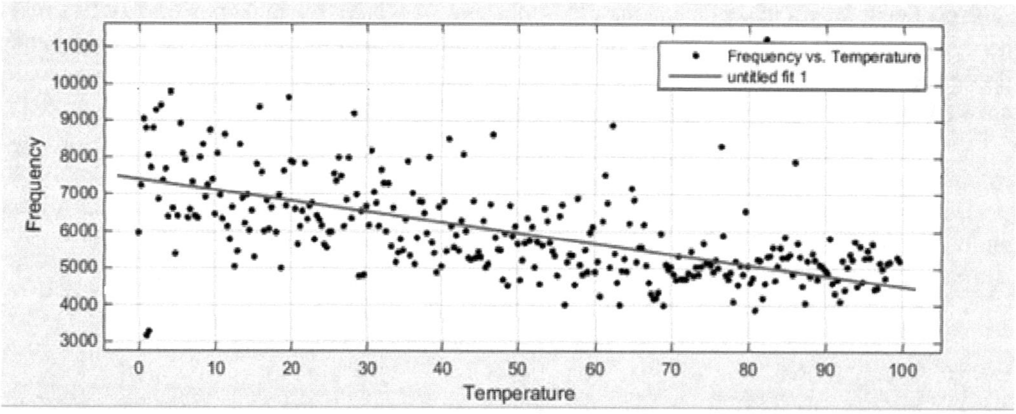


**Figure 10**. Dependence of the wall temperature on the frequency of acoustic oscillations.

## 4. Concluding remarks

The prospect of this development is the possibility of using it in the educational process to demonstrate the processes that arise during a heat exchange crisis, as well as when debugging algorithms and machine learning methods for predicting boiling regimes. The experiment showed that acoustic control systems for the heat transfer crisis for thermophysical systems can be realized, and it is also possible to create algorithms for indirect measurement of the temperature of the shell of heating elements.

## References

[1] Tolubinsky V I 1980 *Heat Transfer during Boiling* (K.: Science Dumka) p 315
[2] Nesis E I 1965 The boiling of liquids *Successes of physical sciences* **87** 615–53
[3] Stepanchenko K P and Belousov P A 2016 Laboratory Scientific-Research Stand for Construction Thermophysical Processes *XII Int. scientific-practical conf. of students, post-graduate students and young specialists* (Obninsk) pp 165–7
[4] Lykov E V and Romanov V V 2011 Acoustic method for diagnosing the liquid boiling crisis *Bulletin of the Voronezh State Technical University* **7** 109–11
[5] Dyakonov V P *Matlab 6.5 SP1 / 7 + Simulink 5/6. Signal Processing and Filter Design* (M .: SOLON-Press) p 576
[6] Pavleyno M A and Romadanov V M 2007 *Spectral Transformations in MATLAB* (St. Petersburg) p 160

# Static strength calculation of DN 350 branch pipe cutting-in in the main circulation pipeline of reactor plant WWER-1000

**A E Ovsenev, N Pimenov, A O Semenov, I S Bibik and M M Balachkov**

National Research Tomsk Polytechnic University, Tomsk, Russia

E-mail: ovsenev1993@mail.ru

**Abstract**. Nowadays the total electricity generated in nuclear power plants in Russian Federation is approximately 18% of all power generation. There are 36 reactors in operation in Russia, 19 of which are reactor plants WWER-1000. A planned operation lifetime of these plants is 30 years and most of them have served for this period. To extend reactor working lifetime it is necessary to carry out the calculation for possibility justification of continued operation of reactor equipment and pipelines. The paper presents the static strength calculation of DN 350 branch pipe cutting-in in the main circulation pipeline of reactor plant WWER-1000. According to calculation results a following conclusion is made: if the static strength of DN 350 branch pipe cutting-in meets the requirements of technical standard, the continued operation of this pipeline section will be possible.

## 1. Introduction

The main circulation pipeline (figure 1) is an element of reactor plant primary circuit designed for a coolant circulation from nuclear reactor to steam generator and backwards. It connects plant main equipment together forming a circulation loop.

The pipeline is made from pipe units connected by welding. The pipe outer diameter is 990 mm, the nominal wall thickness is 70 mm, the basic metal (steel 10GN2MFA) thickness is 65 mm and the platting layer thickness is 5 mm [1].

The main circulation pipeline forms 4 circulation loops, either of which includes:

- A "hot" leg – a pipeline section connecting reactor outlet branch pipe with steam generator outlet branch pipe.
- A "cold" leg consisting of two pipeline sections connecting steam generator outlet branch pipe with main circulation pump outlet branch pipe and main circulation pump outlet branch pipe with reactor outlet branch pipe.

In order to realize the reactor plant technological process the main circulation pipeline is equipped with branch pipes to connect pipelines of normal operation systems, safety systems and impulse lines of control and measuring instruments.

The outer diameter size (850 mm) is chosen according to the conditions of ensuring the acceptable friction loss of main circulation circuit. A "hot" leg of loop No. 4 is connected with pressurizer by the connecting pipeline 426x40 mm. The connecting pipeline inner diameter is chosen according to the limiting conditions of pressure difference acceptable value (0.39 MPa) between the pressurizer and the main circulation pipeline in non-stationary modes without the integrity loss of primary or secondary circuit. A coolant is removed from a "cold" leg of loop No. 1 by the pipeline 219x19 mm to the pressurizer spray [1].

The paper presents the static strength calculation of section of DN 350 branch pipe cutting-in in the main circulation pipeline (figure 2).

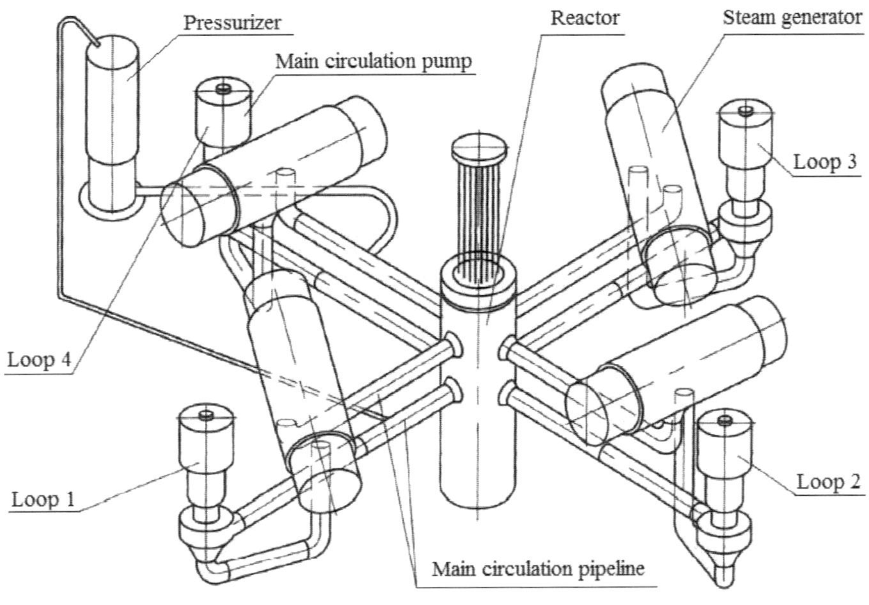


**Figure 1.** The overview diagram of main circulation pipeline.

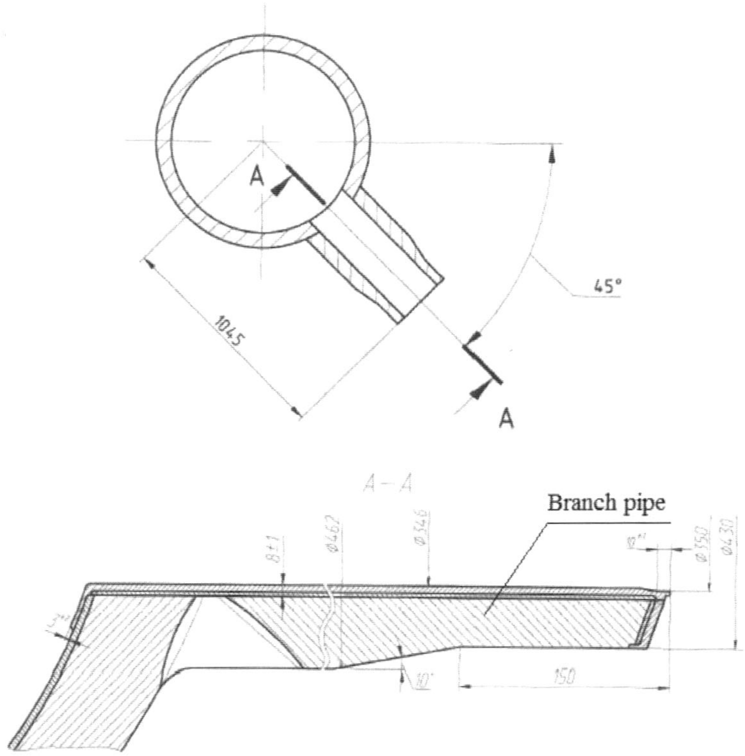


**Figure 2.** The section sketch of DN 350 branch pipe cutting-in in the main circulation pipeline.

IOP Conf. Series: Materials Science and Engineering **487** (2019) 012005 doi:10.1088/1757-899X/487/1/012005

## 2. Operation conditions

At the start of work the main calculation parameters which are necessary for carrying out the strength calculation were chosen [2]. The choice of these parameters is conditioned by the fact that these values characterize all operation modes of nuclear plant. The main calculation parameters are listed in table 1.

**Table 1.** The main calculation parameters.

| Parameter | Value |
|---|---|
| Primary circuit coolant pressure [MPa] | 15.7 |
| Calculation pressure [MPa] | 17.6 |
| Hydrostatic tests pressure [MPa] | 24.5 |
| Hydrostatic tests pressure for a view [MPa] | 19.6 |
| Primary circuit coolant temperature [$^{\circ}$C] | |
| • Reactor inlet temperature | 290.0 |
| • Reactor outlet temperature | 320.3 |
| Calculation temperature [$^{\circ}$C] | 350.0 |
| Hydrostatic tests temperature [$^{\circ}$C] | 135.0 |

In order to determine a fatigue of branch pipe cutting-in material it is necessary to form the list of calculation modes where the material is influenced in a greater degree by the primary circuit coolant during a reactor plant operation.

According to parameter variations in the primary circuit in reactor plant design conditions the modes which describe with desired precision all possible temperature states of considered section are chosen.

The analysis of parameter variations in design conditions showed that in order to justify the section fatigue strength of auxiliary systems cuttings-in in the main circulation pipeline it is enough to consider the modes presented in table 2 [3].

**Table 2.** The forming of calculation modes for cutting-in section of connecting pipeline in the main circulation pipeline.

| Calculation mode | Number of repetitions |
|---|---|
| "Cold" state | – |
| Heating-up to hydrostatic tests temperature | 130 |
| Hydrostatic tests | 130 |
| Planned heating-up from "cold" state | 130 |
| "Hot" state | – |
| Reactor plant power increasing | 733 |
| Reactor plant power decreasing | 200 |
| Planned cooling to "cold" state | 71 |
| Reactor protection system dropping | 100 |
| Steam generator safety valves testing | 640 |
| Main circulation pump shutdown | 120 |
| Nuclear power plant blackout | 5 |
| Closing of monoblock turbine stop valves | 250 |
| Reactor plant accelerated cooling with the speed of 60 [$^{\circ}$C] | 10 |
| Heat transfer tube rupture | 10 |
| Pipelines rupture of primary circuit DN 850 | 6 |
| Steam generator steam line rupture | 16 |
| Prompt main circulation pump jamming | 4 |

## 3. Static strength calculation procedure

The static strength calculation of considered section was carried out using the program complex ANSYS Mechanical [4]. At the calculation start a section finite-element model of connecting pipeline branch pipe cutting-in in the main circulation pipeline was developed (figure 3). To consider a model in more details its enlarged fragment is presented in figure 4.

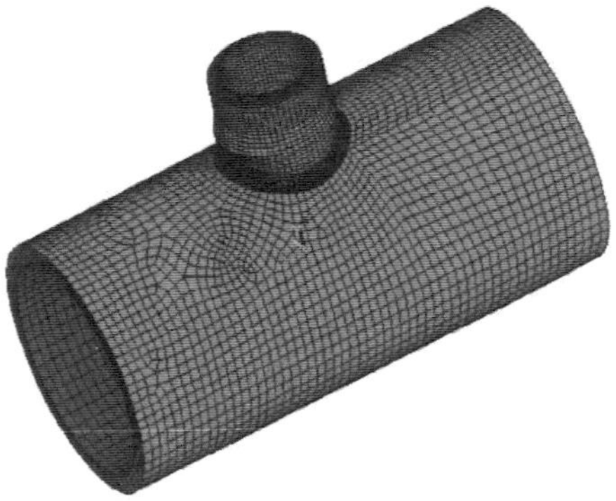

**Figure 3.** The section finite-element model of connecting pipeline branch pipe cutting-in in the main circulation pipeline.

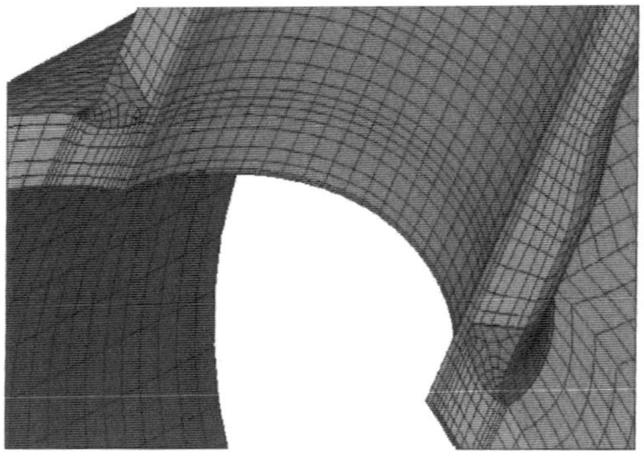

**Figure 4.** The enlarged fragment of section finite-element model of connecting pipeline branch pipe cutting-in in the main circulation pipeline.

During the model developing following features were taken into account: features of geometric design parameters of considered section, features of external forces impact, a physical properties non-uniformity of materials used for section producing and features of temperature influence on cutting-in section elements caused by working media.

A pressure loading is set at inside section surfaces. A tension stress appeared because of closed-loop volume where pressure acts is set at the end surfaces. The stress quantity acting at the end surfaces is calculated using the following formula:

$$\sigma_t = p \cdot \frac{d_{in}^2}{d^2 - d_{in}^2} \qquad [\sigma] \tag{1}$$

где $\sigma_t$ – tension stress [MPa], $p$ – inner pressure [MPa], $d$ – outer diameter [m], $d_{in}$ – inner diameter [m].

During the static strength calculation a fulfillment of strength conditions is being verified as applicable to the considered section under the influence of calculation loads [5]. The stresses determined during the static strength calculation as applicable to the considered section must not exceed the corresponding allowed stresses presented in tables 3 and 4.

The strength calculation including external static and dynamic forces is carried out for the following loading combinations: normal operation, normal operation + design basis earthquake, normal operation + design basis accident + design basis earthquake, normal operation + maximum calculation earthquake.

According to the analysis of stressed states obtained from corresponding loadings impacts by the calculation the most stressed zones of considered sections are chosen (figure 5). Then the corresponding stress categories (general membrane, local membrane, etc.) are calculated in these zones using a stress linearization method.

The element static strength will be regarded as ensured if the calculation reduced stresses of basic stress groups determined from the maximum-shear theory do not exceed the corresponding allowed stresses [6].

**Table 3.** The allowed stress values of different categories.

| Material grade | Calculation case | Calculation stress group | Allowed stresses [MPa] | |
|---|---|---|---|---|
| | | | Definition | Value |
| Steel 10GN2MFA | Normal operation | $(\sigma)_1$ | $[\sigma]$ | 188.5 |
| | | $(\sigma)_2$ | $1.3 \cdot [\sigma]$ | 245.1 |
| | Hydrostatic tests | $(\sigma)_1$ | $1.35 \cdot [\sigma]^{Th}$ | 271.4 |
| | | $(\sigma)_2$ | $1.35 \cdot k \cdot [\sigma]^{Th}$ | 341.7 |

$(\sigma)_1$ – Reduced stresses determined from components of general membrane stresses [MPa].

$(\sigma)_2$ – Reduced stresses determined from components total amount of general or local membrane stresses and general flexural stresses [MPa].

$[\sigma]$ – Nominal allowed stresses [MPa].

$[\sigma]^{Th}$ – Nominal allowed stresses at hydrostatic tests temperature [MPa].

k=1.3 – Coefficient of other geometric form.

**Table 4.** The loading combinations and the allowed stresses.

| Material grade | Calculation case | Calculation stress group | Allowed stresses [MPa] | |
|---|---|---|---|---|
| | | | Definition | Value |
| Steel 10GN2MFA | Normal operation + maximum calculation earthquake, | $(\sigma_s)_1$ | $1.4 \cdot [\sigma]$ | 263.8 |
| | Normal operation breakdown + maximum calculation earthquake, | $(\sigma_s)_2$ | $1.8 \cdot [\sigma]$ | 339.2 |
| | Normal operation + design basis accident + design basis earthquake. Normal operation + design basis earthquake, | $(\sigma_s)_1$ | $1.2 \cdot [\sigma]$ | 271.4 |
| | Normal operation breakdown + design basis earthquake. | $(\sigma_s)_2$ | $1.6 \cdot [\sigma]$ | 341.7 |

$(\sigma_s)_1$ – Reduced stresses determined from components of general membrane stresses including seismic impacts [MPa].

$(\sigma_s)_2$ – Reduced stresses determined from components total amount of general or local membrane stresses and general flexural stresses including seismic impacts [MPa].

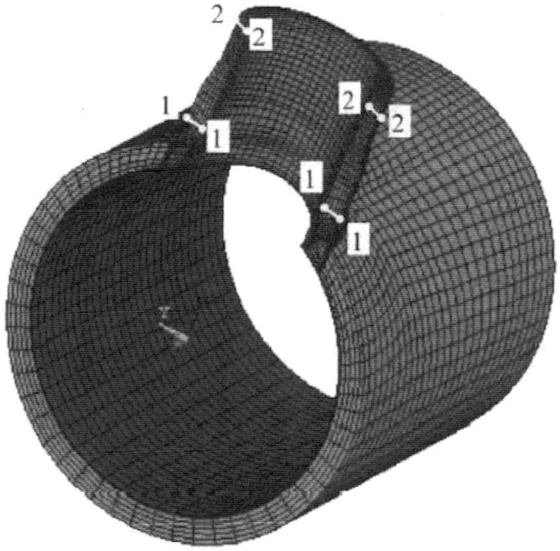


**Figure 5.** The position of reference lines in the section finite-element model of connecting pipeline branch pipe cutting-in in the main circulation pipeline.

## 4. Calculation results

In the course of calculation for normal operation conditions the loading with internal calculation pressure p=17.6 MPa and external loads acting on the branch pipe of connecting pipeline were considered.

The reduced stresses determined from components total amount of local membrane $(\sigma_{\mathrm{mL}})$ and general flexural $\left(\sigma_f\right)$ stresses are shown below while the mechanical loads act on the reference lines 1-1 and 2-2:

$$\left(\sigma\right)_2 = 151.1\, MPa < 1.3 \cdot \left[\sigma\right] = 245.1\, MPa;$$

$$\left(\sigma\right)_2 = 240.9\, MPa < 1.3 \cdot \left[\sigma\right] = 245.1\, MPa.$$

Then the internal pressure loading of hydrostatic test and external mechanical loads from connected pipeline and main circulation pipeline were considered.

The reduced stresses determined from components total amount of local membrane $(\sigma_{\mathrm{mL}})$ and general flexural $\left(\sigma_f\right)$ stresses are shown below while the mechanical loads act on the reference lines 1-1 and 2-2:

$$\left(\sigma\right)_2 = 172.8\, MPa < 1.7 \cdot \left[\sigma\right]^{Th} = 341.7\, MPa;$$

$$\left(\sigma\right)_2 = 340.6\, MPa < 1.7 \cdot \left[\sigma\right]^{Th} = 341.7\, MPa.$$

The loading combination of normal operation and design basis earthquake was considered including loads from external dynamic forces.

The reduced stresses determined from components total amount of local membrane and general flexural stresses during the loading combination of normal operation and design basis earthquake at the reference lines 1-1 and 2-2 are given below:

$$\left(\sigma_S\right)_2 = 155.9\, MPa < 1.6 \cdot \left[\sigma\right] = 301.5\, MPa;$$

$$\left(\sigma_S\right)_2 = 254.5\,MPa < 1.6 \cdot \left[\sigma\right] = 301.5\,MPa$$

The loading combination of normal operation and maximum calculation earthquake was considered including loads from external dynamic forces.

The reduced stresses determined from components total amount of local membrane and general flexural stresses during the loading combination of normal operation and maximum calculation earthquake at the reference lines 1-1 and 2-2 are given below:

$$\left(\sigma_S\right)_2 = 165.4\,MPa < 1.8 \cdot \left[\sigma\right] = 339.2\,MPa;$$

$$\left(\sigma_S\right)_2 = 257.8\,MPa < 1.8 \cdot \left[\sigma\right] = 339.2\,MPa.$$

The loading combination of normal operation breakdown and design basis accident with design basis earthquake was considered including loads from external dynamic forces.

The reduced stresses determined from components total amount of local membrane and general flexural stresses during the loading combination of normal operation breakdown and design basis accident with design basis earthquake at the reference lines 1-1 and 2-2 are given below:

$$\left(\sigma_S\right)_2 = 162.8\,MPa < 1.8 \cdot \left[\sigma\right] = 339.2\,MPa;$$

$$\left(\sigma_S\right)_2 = 256.2\,MPa < 1.8 \cdot \left[\sigma\right] = 339.2\,MPa.$$

## 5. Conclusions

The paper presents the static loading influence on DN 350 branch pipe cutting-in in the main circulation pipeline of reactor plant WWER-1000. The analysis of calculation results revealed that the reduced stresses in the main circulation pipeline and branch pipe did not exceed the allowed values for the following loading combinations: normal operation breakdown + design basis earthquake, normal operation breakdown + maximum calculation earthquake. Consequently, the static strength conditions of DN 350 branch pipe cutting-in in the main circulation pipeline meet the requirements of technical documentation. Therefore it can be concluded that the continued operation of DN 350 branch pipe cutting-in is regarded as possible.

## References

[1] Margolina B Z, Nikolayeva V A, Yurchenkoa E V, Nikolayev Yu A, Erakb D Yu and Nikolayeva A V 2012 Analysis of embrittlement of WWER-1000 RPV materials *Int. J. Pressure Vessels and Piping* **4** 178–86

[2] 1989 *Static Calculaton Standarts of NPP Equipment and Pipelines. RSNP G-7-002-86* (Moscow, Energoatomizdat)

[3] Vojackova J, Novotny F and Katovsky K 2017 Safety analyses of reactor VVER 1000 *Energy Procedia* **127** 352–9

[4] 2013*Program Complex for PC. Program Complex ANSYS for Thermomechanical Problems Solution by Finite Element Method* (ANSYS Mechanical) v. 14.5, ANSYS, inc

[5] 2012 *Strength Calculation Guideline of Equipment and Pipelines of Reactor Plants HPCR, WWER and EGP in Operation Including Operation Beyond Design Lifetime. GD EO 1.1.2.05.0330-2012* (Moscow)

[6] Bykov M A, Siriapin V N and Kozlachkov A N 2015 Statistic Analysis of Meeting the Safety Requirments by Thermohydraulic Analysis Results of Reactor Plant WWER Accidents *Scientific and Technical Collection 'Problems of Nuclear Science and Engineering' Series 'NPP Safety Control'* p 108 ISSN 2224-0853

# Clarification the reliability of electronic components used in nuclear industry

**D S Samokhin, T N Leonova, M Alslman and A D Vostrilova**

National Nuclear Research University MEPhI, Moscow, Russia

E-mail: DSSamokhin@mephi.ru

**Abstract**. When calculating the reliability indicator of any device , under specified operating condition, usually reference data on the characteristics of the reliability of the elements. Data on the reliability of the groups of electrical-radiation equipment (ERE), used in the design, manufacture and operation of equipment, instruments, devices include information on the mathematical models to calculate (predict) values for the operational intensity of product groups, including information on the storage conditions and information on the reliability groups ERE and the coefficients of the models. This paper considers methods for the refinement of correction factors to estimate the failure rate of ERE under specified conditions. The results of the work allow us to clarify the existing coefficients for the prevention and reduction of the failure rate of ERE, as well as to prevent failures of elements and systems due to personnel errors.

## 1. Introduction

For the use of electrical- radiation equipment (ERE), the effects of environmental conditions have to be considered. The reliability of the equipment can be affected by a number of factors, such as, increasing or decreasing of temperature, sudden fluctuation of temperature, thermal shock, and the humidity. To predict the performance and failure of the component, it is necessary to clarify the reliability of each group of ERE. To do this, experiments and information about the reliability of electrical-radiation devices used in the development, manufacturing and operation of equipment, shall be recorded in reference books.

Information about the reliability of the groups ERE and the coefficients of the models includes

- Values of the basic failure rate of ERE groups;
- Values of the coefficients included in the model of forecasting the operational reliability of ERE, and analytical expressions showing the dependence of these coefficients on the factors taken into account.

The values of the operational failure rate of the majority of ERE groups are calculated by the mathematical model:

$$\lambda_e = \lambda_b * \prod_{i=1}^{n} K_i$$

where $\lambda_b$ – basic failure rate of ERIE groups;

$K_i$ – coefficients that take into account the change in the operational failure rate, depending on various factors;

N – the number of factors taken into account.

The coefficients included in the mathematical models for calculating the failure rate of ERE groups can be divided into two groups:

- The first group of coefficients is common for the models of most classes and groups of equipment and characterizes the modes and conditions of their operation, the quality level of ERE manufacturing;
- The second group of coefficients is included in the models of specific classes (groups) of ERE and characterizes the dependence of the failure rate in the given operating conditions on the structural, functional and technological features of ERE.

Reference data should be constantly updated to better predict the behavior and to have an understanding of the shortcomings of the equipment used to develop the most reliable new devices [1].

Depending on the factors affecting on the equipment, the experimental methods are selected. Climatic tests are conducted according to a certain methodology. In the sequence of operations in the experiment, the mandatory steps must be a pre-exposure, an external inspection of the initial test, a final exposure, and a final external examination [2].

## 2. Materials and methods

The Testing of electronic components (EC) for climatic effects is carried out to test the ability of equipment to perform the required functions, to preserve parameters and / or appearance within the established standards under and after exposure. For the reproducibility of the test results, its complete and accurate description is necessary, excluding any uncertainty of interpretation. Proceeding from this, the standards and technical documentation (STD) has adopted such a sequence of operations (stages) of the EC test for climatic effects:

- Pre-exposure equipment (stabilization of their properties);
- Initial parameter measurements and external inspection;
- Installation of device in the chamber and exposure their to the test mode;
- Extraction from the chambers and exposure to restore the properties of equipment (final stabilization of properties);
- External inspection and final product measurements

Pre-exposure is carried out in order to completely or partially eliminate the effects of exposure to equipment in previous operating conditions. Equipment can withstand, as a rule, under normal climatic conditions. Exposure of equipment, whose measurement results can be significantly influenced by relative humidity, is performed These conditions provide for strict maintenance of the temperature (permissible deviation $\pm$ 1 ° C) at a relative humidity of 73 ... 77%. The duration of the preliminary exposure is determined by the time sufficient to establish the thermal equilibrium of the equipment with the environment. Usually it does not exceed 2 hours. Initial and final measurement equipment recommended at the same temperature and humidity environment. When installing equipment in the chamber, it is necessary to ensure that air is freely circulating between the equipment and the walls of the chamber, as well as between the components themselves. The method of installation and position of the equipment in the test are important for the reproducibility of its results. If several variants of the position of the product are possible during operation, the option that provides the greatest test rigidity should be selected. If during the test the electric load on the EC is not amenable, the components are placed on the grids of nylon threads stretched on the supports.

When tested with an electric load, the product is installed on special boards, devices (cassettes, holders, contacting devices). Metal parts of devices must have anti-corrosion coatings. The time of exposure in the test mode is counted from the moment the mode is set in the chamber. This time at high / low temperature should be sufficient for heating / cooling equipment throughout. External inspection of equipment should be carried out in accordance with the STD.

Climatic tests are carried out not only at the design stage of F EC, but also in mass production for the rejection of potentially unreliable equipment (acceptance tests) and to control the stability of production (periodic tests). Modes and conditions of testing EC set depending on the degree of stiffness, which, in turn, is determined by the conditions of further operation of the EC as part of the system. The equipment are considered to have passed the test, if they during and after it meet the requirements specified in technical specification(TS) for this type of test.

To increase the informativeness and effectiveness of climate testing during the development and manufacture of equipment, it is advisable to conduct them in such a sequence in which each subsequent test increases the impact of the previous one, which could go unnoticed. It is recommended that the so-called normalized sequence of climatic tests, including a test at elevated temperature, a short-term test for moisture resistance in a cyclic mode (the first cycle), tests for the effects of low temperature and atmospheric pressure, the test for moisture resistance in a cyclic mode (the rest of the cycles). In this case, between any of these tests may be a break of no more than 3 days, except for the interval between the tests for moisture resistance and the effect of low temperature, which should not exceed 2 hours. Parameters of equipment are usually measured at the beginning and end of the normalized sequence [3].

## 3. Results and Discussion

Testing the climatic effects of high temperature of the medium are carried out to check the performance of EC and/or to preserve their appearance under the influence of high temperature and after it. There are two methods of testing for elevated temperature: thermal and combined load (thermal and electrical). According to the first method, non-heat-releasing EC are tested, the temperature of which during operation depends only on the ambient temperature; on the second - the heat-generating EC, which are heated in the operating state due to the power released under the action of an electric load.

When tested under combined load, the equipment are placed in a chamber and tested under normal or maximum permissible electrical load corresponding to the upper value of the ambient temperature and set depending on the degree of stiffness specified in table 1.

**Table 1.** The value of the temperature depending on the stringency of tests for testing the impact of elevated temperature.

| Degree of stiffness | I | IV | VI | VII | VIII | IX | X | XI | XII | XIII |
|---|---|---|---|---|---|---|---|---|---|---|
| Temperature, °C | 40 | 50 | 70 | 85 | 100 | 125 | 155 | 200 | 250 | 315 |

Heat chambers are used to test for elevated temperatures.

Tests for the effect of a reduced temperature of the surrounding environment are carried out to check the performance of the EC and to preserve their appearance under the influence of a reduced temperature and after it. During the test, the same parameters are measured as during the test for high temperatures. At the same time, to check the efficiency of the EC, it is provided to withstand them under an electric load at a given temperature. First, the EC parameters are measured under normal conditions, then the equipment are placed in a cold chamber, set the temperature limit according to the data specified in the TS and presented in table 2.

**Table 2.** The temperature value depending on the stiffness of the tests for the tests for the effects of low temperature.

| Degree of stiffness | III | IV | VII | VIII |
|---|---|---|---|---|
| Temperature, °C | -10 | -25 | -4 | -60 |

- Withstand EC at a given temperature for a time, which is selected depending on the degree of test rigidity from the time series of values given in TS, but not less than 24 hours;
- raise the temperature in the chamber to the operating low (in accordance with the TS for the tested EC) temperature and withstand them under electric load for a time sufficient to equalize the temperature throughout the volume;
- include EC, conduct measurement specified in the PI and the parameters and verify the operation of the setting elements and switching devices;
- raise the temperature in the chamber to normal and withstand EC for the time required for heating them throughout the volume;
- open the camera, perform external inspection and measurement of EC [4].

Heat and cold chambers are used to test for elevated temperatures. The test for the effect of changes in the temperature of the surrounding environment is carried out to check the performance and maintain the appearance of the EC after the specified impact. Depending on the purpose and operating conditions, as well as on the design features of the EC, the test is carried out by the method of either two chambers (for EC, which under operating conditions are subjected to a rapid change in the temperature of the surrounding environment) or one chamber (with a slow change in the temperature of the medium). For the test, three cycles shall be established, unless otherwise specified in the TS. Each cycle consists of two stages-tests: at low and high temperature of the surrounding.

Thermal Cycling — one of the toughest types of climatic tests-allows to identify hidden design defects and violations of technology.

The test for a thermal shock (on the stability of an ES to the effect of sharp temperature changes) is sometimes carried out along with a test for the effect of a change in the temperature of the surrounding environment.

This test is very rigid and is used to establish the limit strength properties of the EC; its results can be relatively quickly obtaining data on the weakest parts of the structure of the EC. The test is carried out using the method of two liquid baths. In this case, the EC is usually exposed to 10 cycles, unless another number is specified in the TS. The test for one cycle is carried out in two baths: in one, the water has a lowered temperature, in the other, it is elevated. The temperature values correspond to the test modes.

Tests for exposure to high humidity are conducted to establish the moisture resistance of the EC. There are two types of test for moisture resistance: prolonged and accelerated. Long-term testing is carried out to determine the ability of products to maintain their parameters during long-term exposure to humidity and after its end; accelerated testing - with the purpose of prompt detection of gross technological defects in batch production and defects that could arise in previous tests. Both types of tests can be carried out in cyclic (with condensation of moisture) and continuous (without condensation of moisture) modes.

The cyclic mode of the test is characterized by the effect of high humidity in a cyclic change in the air temperature in the chamber. Usually it is used for equipment that do not have compacted casings of all classes, which must remain operational in conditions of dew.

The EC test at electrical load is provided if exposure to humidity in the operating conditions of the equipment under voltage can lead to electrochemical corrosion. As a load in this test is the voltage that provides minimal heat in the test equipment. In most cases, the moisture resistance test is carried out without electrical load.

Product parameters are usually measured at the end of the test (in the last cycle at the end of the last hour of exposure at the upper temperature), without removing them from the humidity chamber.

Heat and humidity chambers are used to test for the effect of elevated temperature [5].

The specific test mode is set depending on the purpose and operating conditions of the EC in accordance with table 3.

**Table 3.** The degree of rigidity of the moisture resistance test, depending on the operating conditions of the equipment during the year.

| Degree of stiffness | Test moods | | | Operating conditions | | |
|---|---|---|---|---|---|---|
| | Relative humidity (high value) (%) | Test Temperature (°C) | Presence of moisture condensation | Relative humidity (average monthly value) (%) | Ambient temperature (°C) | Duration of exposure during the year (months) |
| I | 80 | 25 | - | 65 | 20 | 12 |
| II, III | 98 | 25 | - | 80 | 20 | 2 |
| IV | 100 | 25 | - | 80 | 20 | 6 |
| V | 100 | 25 | + | 90 | 20 | 12 |
| VI, VII | 98 | 35 | - | 80 | 27 | 3 |
| VI, II | 100 | 35 | + | 90 | 27 | 12 |

## 4. Conclusion

To clarify the validity coefficients for the electronic components used at the facilities of nuclear technology, it is necessary to periodically test the climate impacts.

The results of these tests will help to reduce the number of failures of ERE, to assume the time of failure of already used equipment, as well as to bring to zero the number of failures for newly created equipment and to improve the reliability of equipment, which contributes to better safety and economic benefits.

## References

[1] Samokhin D S, Alslman M, Vostrilova A D and Kochnov O Yu 2017 *Experiment for Justification the Reliability of Passive Safety System in NPP* (KnE Life Sciences: in XIII International Youth Scientific and Practical Conference "FUTURE OF ATOMIC ENERGY – AtomFuture 2017") pp 1–6

[2] 22nd Central Research and Testing Institute of the Ministry of Defense of the Russian Federation 2006 *Handbook Reliability of Electronic Radio Products* (Mytishchi: Moscow Region of the Russian Federation) pp 7–8

[3] Spång K and Ståhl G 2002 *Qualification of Electrical Components in Nuclear Power Plants* (Sweden: Management of Ageing) p 49

[4] Gludkin O P 1991 *Methods and Devices for Testing RES and EMU* (Moscow: High School) pp 160–77

[5] Component Technical Committee 2010 *Stress Test Qualification for Passive Components* (USA: Automotive Electronics Council) pp 6–9

IOP Conf. Series: Materials Science and Engineering **487** (2019) 012007 doi:10.1088/1757-899X/487/1/012007

# Influence of boron enrichment in control rods on neutron-physical characteristics of the core of the reactor VVER

**A M Terekhova, Y V Levchenko, T V Matveeva, I S Borisov and I A Chelmakov**

Department of Nuclear Physics and Technology, Obninsk Institute for Nuclear Power Engineering – NRNU MEPhI, Obninsk, Russia

E-mail: YVLevchenko@mephi.ru

**Abstract**. Boron is widely used in nuclear power as a neutron absorber material, thereby creating the possibility of controlling a nuclear reactor by changing the neutron multiplication factor. Natural boron contains about 20% $^{10}B$ and 80% $^{11}B$ with absorption sections respectively 3840 barn and 0.05 barn. The paper deals with changes in neutron-physical parameters of the reactor core of VVER type when the isotope content of $^{10}B$ in boron carbide of absorber rods (AR) of the control and protection system (CPS) and in boric acid solution in the coolant of the VVER reactor is varied, and also analyzes the practicality of increasing the enrichment of $^{10}B$ isotope in the control rods (CR). The model of VVER-1200 fuel assembly (FA) was used for calculations. In the course of work were modelled and the calculations of the three models of the coolant with different boric acid concentrations, as well as two models of AS with different enrichment in the isotope $^{10}B$.

## 1. Introduction

At present, boron carbide is the most widely used as an absorbing material in the control rods of CPS. The choice of this material is due to its high neutron absorption efficiency in a wide range of energies, high melting point, manufacturability and relatively low cost [1]. The most important characteristic of boron carbide is the thermal conductivity, which determines the temperature conditions in which the cores of the absorbing elements in the reactor work. It averages 22-33.5 W/(m·°C) depending on density [1]. The coefficient of thermal conductivity has a strong dependence on the content in the sample of the isotope $^{10}B$.

The chemical stability of boron carbide is explained by strong covalent bonds between carbon and boron atoms and a low concentration of non-localized electrons. Boron carbide well resists the action of mineral boiling acids.

As a result of neutron absorption by isotope $^{10}B$, α-particles with high kinetic energy and low path length are obtained, so when increasing the enrichment of boron, it should also be borne in mind that:

1) absorbers made of boron or its compounds are heated and require their cooling;

2) α-particles during prolonged irradiation of boron by neutrons violate the structure of the material, which leads to the deformation of construction made of it [2].

Typically, $B_4C$ is compressed into tablets.

The compatibility of boron carbide with structural materials is considered to be the main criteria for the efficiency of absorbing elements. As materials of shells of absorbing elements, stainless steels and alloys, as well as zirconium alloys, are most often used. In the control rods of VVER type nuclear reactors, boron carbide powder is widely used as the core of absorbing elements, which is in direct contact with a shell of steel [3].

In power reactors of high power, unevenness in the core is observed, therefore the control rod systems are cumbersome, which affects the cost of the control system and the reliability of its operation. There are several reasons for using absorber materials dissolved in the coolant, in particular, reducing the number of absorbing rods along with the drives and electrical equipment of the control system, which saves cost.

Of the potentially useful elements (boron, cadmium and gadolinium), boron compounds, although they have the smallest neutron absorption cross section, however, their physico-chemical properties (solubility and resistance) are most preferable under the conditions of the first circuit. Boric acid is evenly distributed over the fuel assemblies, and therefore, when the concentration is changed, the distribution of energy release in the core is not impaired. It was also found that the concentration of boric acid with alkaline additives (potassium, $NH_4$ and lithium) does not significantly increase the corrosion rate of the contour materials in the water of the primary circuit in the reactor core. The presence of boric acid does not cause special complications in the purification of the water of the primary circuit, including the boric acid itself. In this regard, water regimes of the first circuit using boric acid have found wide application both in our country and abroad. If the absorbing rods are used to react quickly to the requirements of reactor operation (power changes, stop and start), the control of the boric acid concentration is maintained in the critical state during slow transient processes (burnout, xenon poisoning).

The concentration of boric acid varies over time in accordance with the requirements of operation, which occurs remotely and automatically at the command of the reactor operator.

The aim of this work is to study the influence of the degree of enrichment of boron in the boron carbide of CR and the boric acid in the coolant on neutron-physical characteristics (NFC) of core of reactor VVER.

## 2. Materials and methods

For calculations in the work, the Code UNK software complex was used. The basic programs of the Code UNK package are oriented to perform calculations with accuracy limited only by the capabilities of computer technology, that is, the ability to perform calculations with a large number of energy groups, with a large number of spatial zones. Unlike conventional engineering programs, the mathematical models of physical processes realized in the programs of the Code UNK package have minimal simplifications of real processes. In fact, there are two of them and both concern the UNKCELL program. The first approximation is that when calculating the neutron spectrum in a cell or reactor cassette, an approximation is used for isotropic neutron scattering and isotropic neutron sources. Second, the isotope cross sections in the region of unresolvable resonances are determined within the framework of the group approximation with standard blocking by the equivalence theorem.

Small approximations of real physical processes and the fact that all programs are developed on the basis of deterministic techniques and algorithms allow the Code UNK software package to be classified as a package of precision deterministic programs. This means that the accuracy of calculating the various characteristics of the reactor, subject to the saturation of the spatial and energy grids, will correspond to the accuracy of statistical calculations based on the Monte Carlo methods.

The Code UNK program has its own nuclear data library completely generated from the files of the estimated nuclear data ENDF/B-6, JEF-2.2, JENDL. In the version of the program used in the program, the library contains 337 isotopes. The main program library contains 89 groups of isotope sections (24 groups in the deceleration region from 14.5 MeV to 2.15 eV and 65 thermal groups from 1 to 2.15 eV) [5].

A special feature of the UNKCELL program is the detailed calculation of the neutron spectrum in the resonance energy region. This is achieved by performing calculations in a large number of energy groups (about 7000). For these calculations, the nuclear data library of resonant isotopes contains group sections in the fine-group partition. This library is also derived from the ENDF/B-6 estimated nuclear data files.

The spatial distribution of the neutron field in the cell is calculated by the method of probabilities of the first collisions. Analytic formulas for calculating probabilities are used for simple geometries. For complex two-dimensional geometries, the probabilities of the first collisions are calculated numerically.

The UNKCELL program consists of three main program modules:

MACSEC – calculation of macroscopic cross section of the cell;

CELLHI – calculation of the spatial energy distribution of neutrons in a cell or cassette;

AVR – averaging program, in the calculated cell spectrum, group constants and functionals.

All UNKCELL program modules are written in the algorithmic language FORTRAN-77.

The calculation of the cell or cassette of the reactor involves the execution of three consecutive stages:

1) Calculation of macroscopic cross section of various materials that make up the cell. At this stage, the corresponding values are selected from the nuclear data library of different isotopes as a function of temperature and the dilution cross-section calculated by the program. A macroscopic cross sections library is being created. This part of the calculation is performed by the MACSEC program.

2) Direct calculation of the spatial energy distribution of neutrons in the cells and calculation of the effective multiplication factor. In accordance with the generated task, the calculation is carried out either within the framework of the 89 group approximation or in 7000 energy groups. In the first case, resonance blocking is carried out within the framework of the approximation of the equivalence theorem, in the second case, a detailed calculation of the neutron spectrum in the resonance energy region is carried out. This part of the calculation is performed by the CELLHI program.

3) The calculation results are processed on the basis of the obtained neutron field distribution in the cell. The averaged characteristics can be calculated as the average over the entire cell volume, and by its individual zones. In calculating the average characteristics of the cell, the following macroscopic cross sections and functionals are calculated in a given number of energy groups: the absorption, fission, elastic scattering, inelastic scattering cross sections, the mean cosine of the scattering angle, the cross-section matrix of the intergroup transitions in the deceleration region and the thermalization region. The value of $k_{eff}$ is calculated as the ratio of the integral over the fission cell to the absorption. Calculation of the microscopic cross sections of absorption and fission of various isotopes that make up the cell for the purpose of their subsequent use for calculation of fuel burnup is carried out. This stage of calculation is performed by the AVR program.

To perform the calculations, a model of the VVER-1200 reactor fuel assembly. A fuel element with a fuel pellet diameter of 7.60 mm and a diameter of a central hole in a fuel pellet of 1.2 mm was used as the fuel element. Enrichment of fuel is limited to 4.95%. In the design of the VVER-1200 reactor, to reduce the concentration of boric acid in the coolant and to equalize the power density field, part of the fuel assemblies integrated with the fuel burned up the absorber in the form of gadolinium oxide ($Gd_2O_3$). But the work will use fuel assemblies without the gadolinium fuel rod, since the aim of the work is to study the effect of boron on the neutron-physical parameters of the core.

Figure 1 shows the VVER-1200 core model.

The main technical characteristics of fuel assemblies are presented in table 1. Data are taken from source [7].

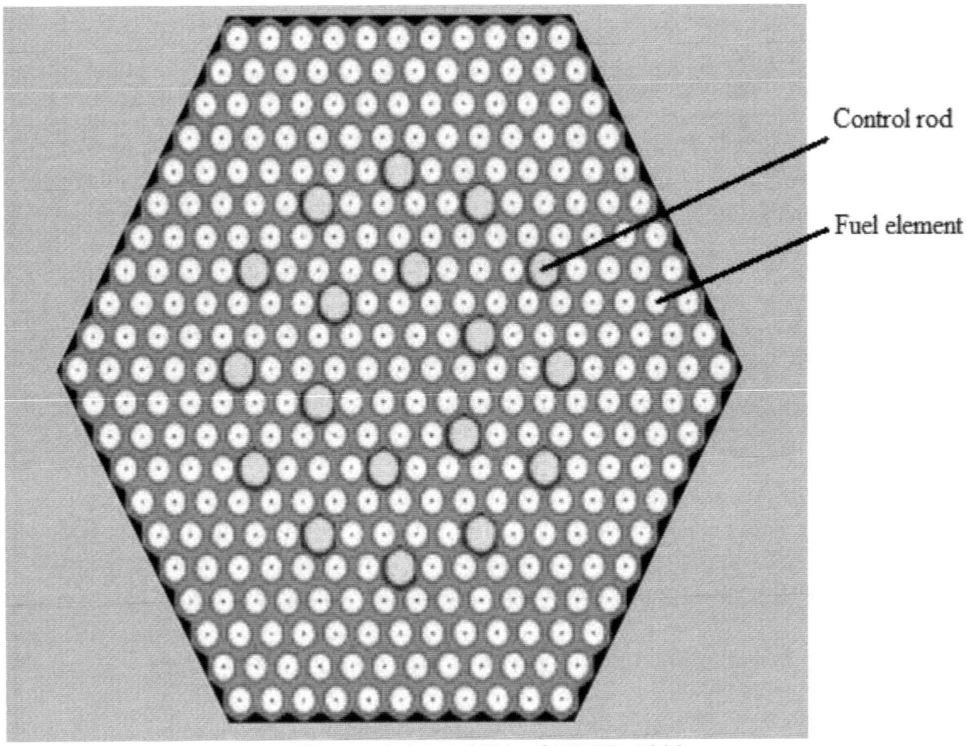


**Figure 1.** Model FA of VVER-1200.

**Table 1.** Main technical characteristics of VVER-1200 FA.

| Characteristic | Value |
|---|---|
| Shape of fuel assemblies | Hexagonal prism |
| Height of FA (mm) | 4570 |
| FA size «flat-to-flat» (mm) | 235 |
| Number of fuel rods in FA (pieces) | 312 |
| Step between fuel elements (mm) | 12,75 |
| Fuel | $UO_2$ |
| Fuel density (kg/m$^3$) | $10.4 \cdot 10^3$ |
| Cladding | Alloy Zr+1%Nb |
| The outer diameter of the cladding (m) | $9.1 \cdot 10^{-3}$ |
| The inner diameter of the cladding (m) | $7.72 \cdot 10^{-3}$ |
| The height of a column of fuel in cold condition (m) | 3,73 |
| The outer diameter of the fuel pellet (m) | $7.57 \cdot 10^{-3}$ |
| Diameter of the central hole in the fuel pellet (m) | $1.5 \cdot 10^{-3}$ |

The content of the $^{10}B$ isotope in the boron of the control and protection system control rod (CPS CR) and in boric acid in the VVER-1200 project is assumed to be equal to the natural (19.8% at.). At the beginning of core life, the critical concentration of boric acid in the coolant is 5.39 g/kg.

In the course of the work, 3 models of a cell of a fuel assembly with different contents of $^{10}B$ in a coolant were considered:

Model 1. Heat carrier without boric acid.

Model 2. The coolant with boric acid with a natural isotope content of $^{10}B$ is 19.8%.

Model 3. The coolant with boric acid with isotope $^{10}B$ content of 90%.

To simplify the simulation, it is considered that the concentration of boric acid does not change during the reactor life time. This does not make a big contribution when comparing models.

To analyze the effect of the degree of boron enrichment in the boron carbide of the neutron-absorbing rod system on the NFC of the fuel, two fuel assemblies of the VVER-1200 reactor were modeled:

1. Using a CPS CR with boron carbide with natural enrichment of $^{10}B$.
2. Using a CPS CR with boron carbide with an enrichment of 90% to $^{10}B$.

## 3. Results and Discussion

Calculations of NFC of the core in the software-computing complex Code UNK for boron carbide with natural enrichment and enrichment of 90% in $^{10}B$ isotope were performed. The density of boron carbide $B_4C$ was estimated to be equal to 1.7-1.8 g/cm$^3$ [4]. The dependence of the value of the multiplication factor on enrichment is shown in table 2.

**Table 2.** Values of the multiplication factor depending on the enrichment at the beginning of the fuel campaign.

| The enrichment in the isotope $^{10}B$ | CR in the upper position | CR in the lower position |
|---|---|---|
| Natural | 1.463 | 0.909 |
| 90% | 1.463 | 0.791 |

As a result of the calculations, it can be concluded that the amount of CR can be reduced by 4 practically without changing the NFC of the core while enriching the boron carbide to 90% by $^{10}B$. At the beginning of the fuel campaign with the lowered AR $k_{eff}$=0.909. This, in turn, makes it possible to increase the fuel load at the same core volume, which allows increasing the duration of the reactor operation without overload. However, the enrichment by the isotope $^{10}B$ entails additional capital investments.

When considering three models of FA with different content of $^{10}B$ in boric acid in the coolant, it turned out that the model 3 is not suitable for comparison, as the effective multiplication factor at the beginning of the fuel campaign was 0.896.

The concentration of boric acid (enriched to 90%) in water was calculated, at which the effective multiplication factor at the beginning of the fuel campaign would be equal to the effective multiplication factor at the beginning of the fuel campaign for model 2. The concentration of the boric acid amounted to 1.165 g/kg (model 4). The analysis of the results shows that without changing the NFC of the core can be dosed into the coolant much lower concentration of boric acid in the enrichment of 90% on $^{10}B$.

Enrichment change can affect the value of the temperature coefficient of reactivity at the temperature of the coolant, which, of course, should be taken into account in the design. An increase in the content of boric acid in the coolant makes a positive effect of reactivity at the coolant temperature, as shown in table 3.

**Table 3.** Values of the effective multiplication factor, reactivity and temperature coefficient of reactivity for the coolant temperature for models 1, 2, 4 at the beginning of the fuel campaign.

| Model | Coolant temperature, K | $k_{eff}$ | Reactivity, Δk/k | $\alpha_{Tcoolant}$, 1/K |
|---|---|---|---|---|
| 1 | 584.8 | 1.462731 | 0.31635 | $-0.3 \cdot 10^{-6}$ |
| | 484.8 | 1.463379 | 0.31665 | |
| 2 | 584.8 | 1.275554 | 0.216 | $5.5 \cdot 10^{-6}$ |
| | 484.8 | 1.27465 | 0.21547 | |
| 4 | 584.8 | 1.275365 | 0.2159 | $5.5 \cdot 10^{-6}$ |
| | 484.8 | 1.274454 | 0.21535 | |

The change of the multiplication factor for each model depending on the burnup time is shown in figure 2.

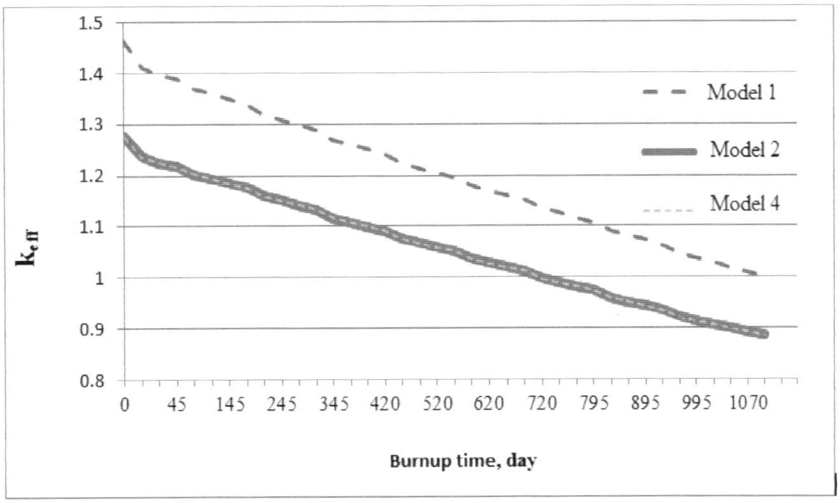


**Figure 2.** The change in the effective multiplication factor from the burnup time in different usage patterns of boron.

It should be noted that when using MOX fuel or REMIX fuel in a nuclear reactor, the absorbing ability of rods with natural boron enrichment by $^{10}B$ isotope will not be enough to compensate for excess reactivity. In this regard, it is necessary to increase the content of $^{10}B$. This means that increasing the enrichment of boron makes sense.

## 4. Conclusion

As a result of calculations it was found that highly enriched boron allows reducing the number of CPS CR. It can be concluded from this that the use of enriched boron carbide in AR makes it possible to load more than 1/3 of the core with fuel assemblies with MOX fuel, and also makes it possible to increase the fuel enrichment to 6% and higher in the isotope $^{235}U$, which is not possible with the regular CR.

The use of an enriched boron isotope of $^{10}B$ in boric acid has both positive and negative sides. The advantages include the possibility of dosing a lower concentration of boric acid into a coolant. Of the minuses can be noted the price increase, which is connected with the necessity of enrichment, it is also theoretically less smooth control of the reactor.

## References

[1] Risovany V D, Zakharov A V and Klochkov E P 2012 *Absorbing Materials and Regulators of Nuclear Reactors* (Moscow: MEI)
[2] Risovany V D, Zakharov A V, Klochkov E P and Guseva T M 2011 *Borov Nuclear Engineering 2e, revised and added* (Dimitrovgrad: JSC 'SRC NIIAR')
[3] Gladyshev A M 1985 *Nuclear-Physical, Chemical Ishmechanic Properties of Boron Carbide: Review. FEI - 0203* (Moscow: TsNIIatominform)
[4] Risovany V D, Varlashova E E and Fridman S R 1998 *Absorbing Materials and Regulators of VVER and PWR Reactors: Status, Problems and Solutions* (Dimitrovgrad: SSC RF NIIAR)
[5] Davidenko V D and Tsibulsky V F 2000 *Calculation of Burnout in the Code UNK Program: Neutronics-1999* (Obninsk)
[6] Volkov Y V 1996 *Physicotechnical Foundations of Designing Nuclear Reactors: Manual on the Course 'Design of Nuclear Reactors* (Obninsk: IATE)
[7] Prytkov A N, Tereshchenko A B, Golubev E I and Boev I A 2017 *Features of the First Nuclear Fuel Loading of the Innovative Power Unit under the Project NPP-2006* (Moscow: Izvestiya Vysshikh Uchebnykh Zavedenii, NRNU MEPhI) pp 121-3
[8] Rules for Nuclear Safety of Nuclear Reactor Units NP-082-07
[9] Ukraintsev V F 2000 *Reactivity Effects in Power Reactors* (Obninsk: IATE)
[10] Vladimirov V I 1986 *Practical Tasks for the Operation of Nuclear Reactors* (Moscow: Energoatomizdat)

IOP Conf. Series: Materials Science and Engineering **487** (2019) 012008 doi:10.1088/1757-899X/487/1/012008

# Influence of operating conditions on the accuracy of the automated system for measuring geometric parameters of graphical methods and measuring deviations of technological channels from the vertical of channel reactors

**D A Teterevenkov, S I Minin and O S Gorlanova**

Department of Nuclear Physics and Technology, Obninsk Institute for Nuclear Power Engineering – NRNU MEPhI, Obninsk, Russia

E-mail: teterevenkovda@gmail.com

**Abstract.** During operation of the reactor, along its active zone, the energy release is unevenly distributed, consequently also the temperature. These operational parameters introduce an error in the measurement of the system. Of practical interest is the study of the temperature stability of ultrasound sensors, as well as the effect of vibration on the oscillations of the ultrasonic signal. The system solves the problem of controlling the curvature and the diameter of the cells of the graphite masonry of the EGP-6 reactors in two mutually perpendicular planes during the routine preventive and overhaul repairs. Diameter control is carried out by means of 4 sensors of displacement of the resistor type fixed on the sensor probe block. Each of the sensors is mechanically connected to the roller, which is in direct contact with the channel wall. Diameter measurement uses data from sensors connected to two opposite rollers. Curvature control is performed using an ultrasonic inclinometer filled with liquid, which is located in the tail part of the probe. The inclinometer is designed for measuring angles of inclination in 2 mutually perpendicular planes. The data from the inclinometer is transmitted via the interface to an analog-to-digital converter, which is located in the measuring unit and is designed to convert the interface to USB.

## 1. Introduction

Diagnostics of graphite blocks and technological channels (TC) by measuring deviations of technological channels from the vertical, as well as monitoring the state of the structural materials of the core of the channel reactors, is a very important task at the NPP. The state of the structural materials largely depends on the safe operation of the nuclear reactor.

Under the influence of radiation exposure and temperature, the physical properties of metal and graphite change during the long-term operation of the channel nuclear reactor. The ducts undergo deformation. The diameter of the channels increases. The swelling of graphite under the action of irradiation leads to a decrease in the diameter of the graphite cell. As a result, the gap between the channel and the graphite cell decreases until the channel is fully compressed and jammed [1].

The monitoring system solves this problem, controlling the curvature and the diameter of the cells of the graphite masonry of the EGP-6 reactors in two mutually perpendicular planes, in the course of carrying out preventive and capital repairs.

## 2. Materials and methods

The technological channels of the reactor are placed in a graphite masonry. The masonry of the reactor has a cylindrical shape with a diameter of 6 m and a height of 5.25 m. It is composed of individual graphite and in the upper part of cast-iron blocks of square section. The central part of the graphite masonry with a diameter of 4.1 m and a height of 3 m, which is the active zone of the reactor, consists of 333 vertical columns with holes of 88.6 mm in diameter along the entire height, which accommodate 273 working channels and 60 CPS channels. The masonry of the reactor is enclosed in a cylindrical hermetic casing. The cell diameter of the graphite masonry of the EGP-6 reactor is 85 - 95 mm. Between the channel and the walls of the graphite cell spring graphite rings are placed split. One part of rings adjoins the channel walls (small rings) and the other to the walls of the graphite cell (large rings) (figure 1).

As a result, a gap is created to allow the channel to be shifted within certain limits [2].

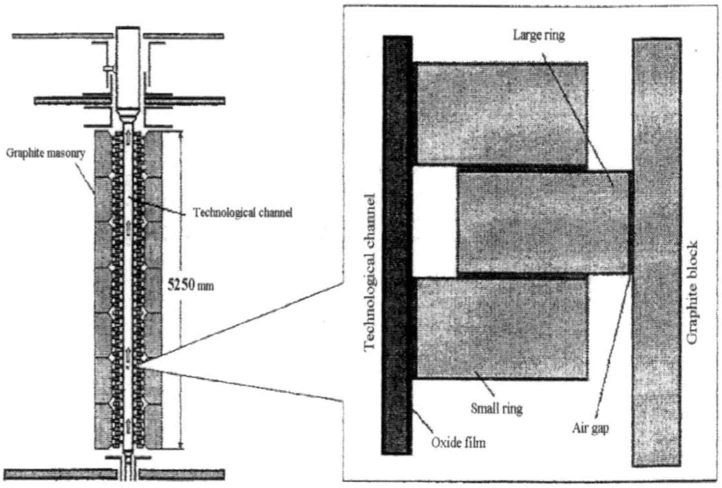


**Figure 1.** Scheme of the technological system 'channel-graphite'.

## 3. Results and Discussion

### *3.1. Temperature distribution in the active zone RBMK-1000*

The temperature regime during the measurement is set by the temperature of the coolant circulating in the TC during the shutdown period of the reactor between the pressure head and the collection manifold. As a rule, the difference in the temperatures of the cooling water at the inlet and outlet of the TC is 30–40 °C, due to sufficient energy release in the core of the reactor. With a low-intensity mode of feeding the TC with cooling water, its temperature can reach 90–100 °C.

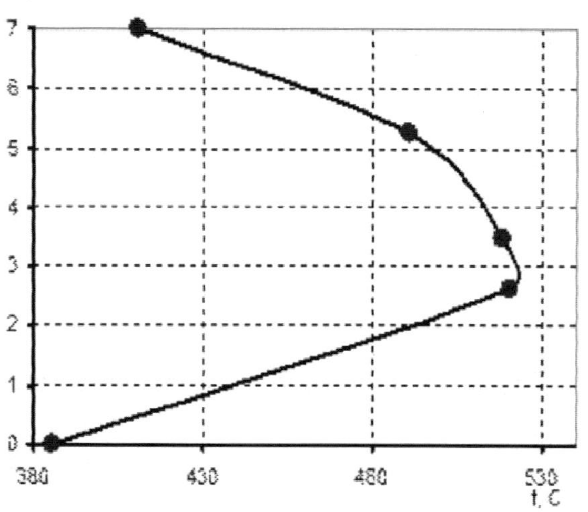


**Figure 2.** Graph of temperature change in graphite masonry by channel height.

Thus, due to the uneven distribution of energy release, consequently the temperature in the core, the study of the temperature stability of ultrasonic sensors is of practical interest (figure 2).

In order to investigate the influence of temperature and vibration on the indications of diagnostic equipment, laboratory tests were conducted measuring angles at a certain level of vibration and various temperatures.

*3.2. Technique for measuring the angle of deflection of the technological channel*

The spatial position of the path of the technological channel is determined by the angle of inclination (zenith angle), the azimuth angle, the distance from the zero mark of the channel to the point of measurement of the angles (figure 3). The angle of inclination is the angle between the TC axis or the tangent to it and the horizontal, and the zenith angles , 1 – are the angles between the same axis or the tangent to it and the vertical. The sum of these angles is always 90, and they lie in a vertical plane, called zenith or apsidal. Based on this technique, the curvature of the technological channel was calculated [3].

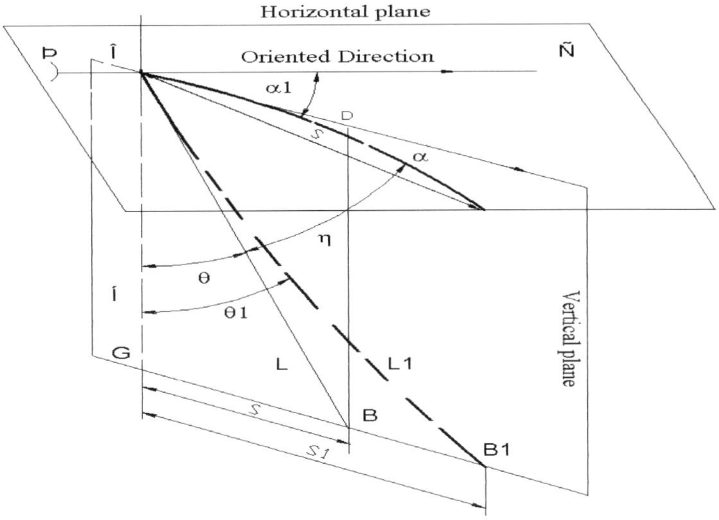


**Figure 3.** Elements defining spatialaxial position of the process channel.

For the experimental confirmation of the theory, a piezoelectric ultrasonic transducer of the angle of inclination was used, made in the form of a glass with liquid and an ultrasonic sensor in the form of a block of ultrasonic sensors. The instruments were filled with water and transformer oil. To study their behavior in measuring devices, experimental measurements were made. In the course of the measurements, information was used on the goniometers, and it was fully used for experimental confirmation of the theory [4].

The signal from the ultrasonic sensor (or inclinometer) was transmitted to the ultrasonic flaw detector "UD2-12", the analog signal with "UD2-12" was transmitted to an analog-to-digital converter, the digital signal was transmitted to a computer on which the program "PicoScope" plotted diagrams (figure 4), at which the arrival time of the signal could determine the angle of inclination.

The following angles were measured: the sensitivity angle of the sensors, the maximum angle of inclination. Angle of sensitivity was measured during vibration as well. An experiment was also carried out with liquid poured into the sensors.

In the course of the work done with the ultrasonic sensor and inclinometer, it was found that for diagnostics of the graphite masonry of the EGP-6 reactor and for the diagnostics of other reactor installations, it is appropriate to use transformer oil as a liquid filled in diagnostic equipment. Due to its viscosity, and hence the resistance to vibration generated by the operation of the reactor, there will be less interference in the signal from the diagnostic equipment, which makes control difficult. As a result, it is preferable to use transformer oil [5].

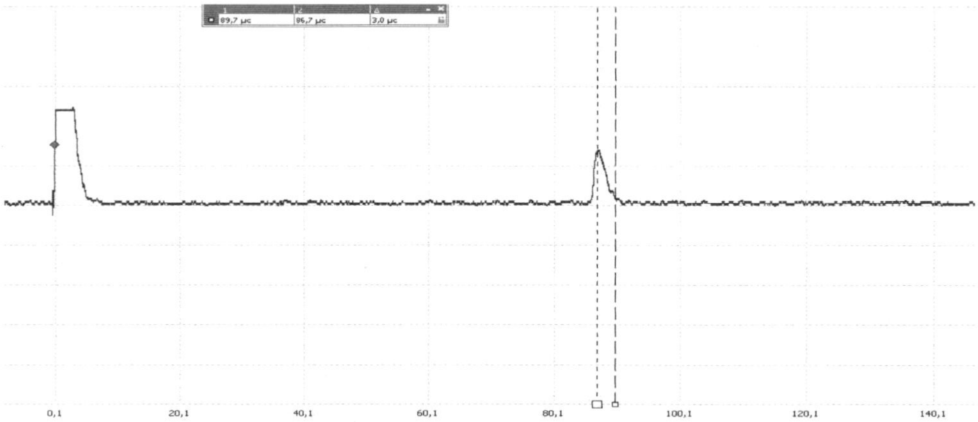


**Figure 4.** Diagrams obtained from the "PicoScope" program.

In the course of experimental work on measuring the angles of the sensors, changing the temperature of the filled liquid, we can conclude. As the temperature of the transformer oil increases, the ultrasonic wave velocity decreases, so the arrival time of the ultrasonic signal to the receiver increases, so the angle of inclination increases. And with increasing water temperature, the speed of the ultrasonic wave, strangely enough, increases, so the arrival time of the ultrasonic signal on the receiver decreases, and therefore the angle of inclination decreases. This is explained by the fact that with increasing temperature, the compressibility is mainly changed. In all liquids other than water, with increasing temperature, the compressibility increases and the ultrasound velocity decreases.

Among the main temperature effects affecting the accuracy of the goniometers, the effect is the change in the velocity of an ultrasonic wave in a liquid with increasing temperature. The speed of ultrasound in liquids is determined by the adiabatic compressibility βad and density ρ:

$$C = \sqrt{\frac{1}{\rho \times \beta}} \quad (1)$$

The analytical expression for the velocity C of ultrasound as a function of temperature t has the form:

$$C = Co + b\,(t - to) \quad (2)$$

For all liquids other than water, the temperature coefficient b is negative. The compressibility of water decreases with increasing temperature and the ultrasound velocity increases.

Therefore, for more accurate measurements, it is necessary to constantly monitor the operating temperature of the fluid in the diagnostic equipment and, at a given temperature of the liquid medium, refer to laboratory studies of the effect of temperature on the measured angles. This requires a temperature sensor in the equipment.

### *3.3. Automated system*

The automated control system consists of functionally completed blocks, communication between which is carried out through detachable connections. The probe with sensor blocks is equipped with upper and lower centering wheels. The lower an inclinometer centering wheels are connected to the linear displacement sensors. In the upper part of the probe is located. The probe through the swivel joint is fixed to the "boom" of the crane beam (figure 5).

To the probe, through communication lines, the electronic block of registration and information processing is connected. The electronic unit for recording and processing information is connected to a computer.

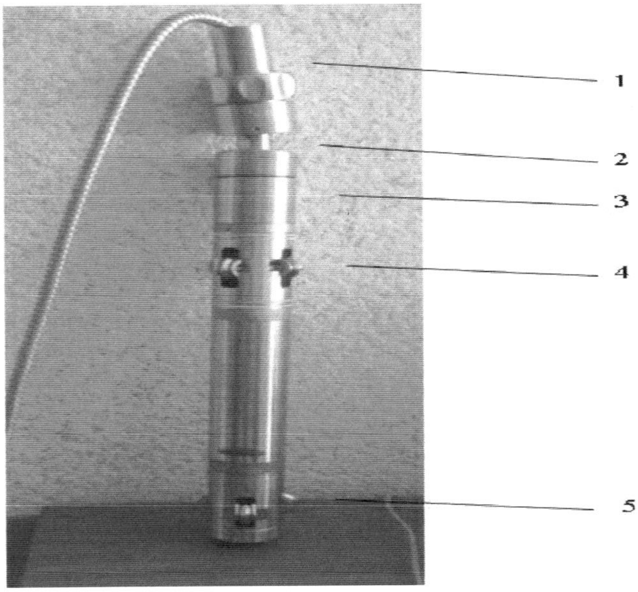


**Figure 5.** Probe with sensor blocks. 1 – fixation block to the "rod", 2 – articulated joint, 3 – section inclinometer, 4 – upper centering wheels, 5 – bottom centering wheels.

Diameter control is carried out by means of 4 sensors of displacement of the resistor type fixed on the sensor probe block. Each of the sensors is mechanically connected to the roller, which is in direct contact with the channel wall. Diameter measurement uses data from sensors connected to two opposite rollers.

Curvature control is performed using an ultrasonic inclinometer filled with liquid (figure 6), which is located in the tail part of the probe. The inclinometer is designed for measuring angles of inclination in 2 mutually perpendicular planes. The data from the inclinometer is transmitted via the interface to an analog-to-digital converter, which is located in the measuring unit and is designed to convert the interface to USB.

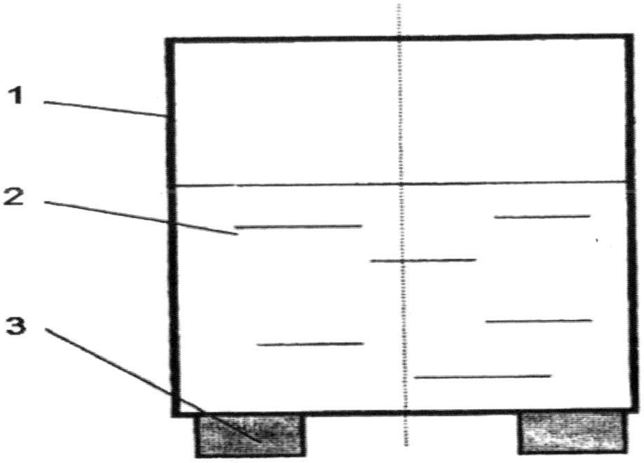


**Figure 6.** Ultrasonic inclinometer filled with liquid.
1-body, 2-liquid, 3-piezoelements.

To carry out work to control the geometric parameters of the cells of the graphite masonry, the probe that is part of the system must be moved along the axis of the cells of the graphite masonry by means of a rod suspended from the crane beam or by a launching device [3].The control of the parameters of the cells of the graphite masonry is carried out with contact method, without using of liquid. The software provides the processing of information for controlling the parameters of the cells of the graphite masonry (diameter and curvature). Input of data goes in a dialogue with the operator through the video terminal of the computer. The result of the input is the system data file: a file that is input information for the program for reading and adjusting data.

## 4. Conclusion

In this paper, the influence of operating conditions on the accuracy of the automated system for measuring geometric parameters of graphite masonry and measuring the deviations of technological channels from the vertical of channel reactors was investigated. During the study, the angles were measured at different temperatures of the filled liquid in the equipment of the automated system and the influence of vibration on the measurements was investigated. This will increase the safety of nuclear power plants by measuring the geometric parameters of the graphite masonry and the deviations of the technological channels without significant interference in the equipment. Thanks to the diagnosis of the canal and masonry, NPP personnel will have an accurate picture of the state of the core, minimally exposing itself to irradiation.

The paper presents the results of experimental studies that were carried out at the stand with the help of experimental equipment (ultrasonic sensor, inclinometer, flaw detector "UD2-12", analog-digital converter, computer, optical quadrant, calipers). The behavior of transformer oil and water in diagnostic equipment at certain temperatures has been studied. Measuring the angles of the sensors relative to the level, conclusions are drawn on their behavior in the TC.

## References

[1] Trofimov A I, Minin S I, Kalenishin M G and Nakhabov A V 2008 Automated control system for geometric parameters of graphite cells of EGP-6 reactors *Nuclear Engineering* **3** 123–7

[2] Trofimov A I 1994 *Ultrasonic Monitoring Systems of Curvature of Technological Channels of Nuclear Reactors* (M .: Energoatomizdat, 1994) pp 230–5

[3] Trofimov A I, Popov V V and Torgashov A L 2004 *Device for Measuring Deviations of the Object from the Vertical* (M: Izd. MSTU N E Bauman) pp 341–57

[4] Trofimov A I, Ermolaev P N and Tkachenko V I 1996 *Device for Measuring the Curvature of Technological Channels of Nuclear Reactors* (M.: Energoatomizdat) pp 156–69

[5] Trofimov A I and Popov V V 2008 Ultrasonic tilt indicator *Nuclear Engineering* **3** 242–54

# Application of simulation modelling approaches for analyzing γ-radiation characteristics of a plume induced by a Nuclear Accident at NPP

**A I Ksenofontov, A P Elokhin and E A Alalem**

National Research Nuclear University-Moscow Engineering Physics Institute (NRNU-MEPhI), Moscow, Russia

E-mail: issaalem2@yahoo.com

**Abstract**. Method Monte Carlo is applied to assess the radiological impacts to the environment. As such cases, determining the dose rates due to external radiation source is considered, that is induced by radioactive inert gases when they are released from a nuclear power plant; also to estimate the volumetric concentrations of the released radioactive inert gases, and other characteristics associated with the use of gamma detectors of the automated radiation monitoring systems of the environment and unmanned aircraft radiometric system. In order to assure the reliability and the accuracy of the calculations by Monte Carlo method for the dose rates estimations, the results are compared with the results that obtained by the integral method, which showed satisfactory agreement.

## 1. Introduction

The calculated values of the radiation characteristics of radioactive pollutant, which is propagating in the atmosphere as a result of radiation accidents at nuclear power plants or other nuclear facilities are requiring both the meteorological parameters of the atmosphere that characterizing its stability state, functional dependence of volumetric activity on these parameters, and the radiation characteristics of the pollutant itself, determined by the radionuclide composition, the total activity and the γ-radiation energy.

In work [1] to determine the functional dependence of the volumetric activity of the radioactive pollutant on the coordinates $q(x, y, z)$ and the meteorological parameters of the atmosphere in the framework of the near surface layer model [2, 3], the meteorological characteristics of the specified atmospheric layer were measured, Including (surface wind velocity $u(z)$, the turbulent diffusion coefficient $k(z)$, and the turbulent pulsation energy $b(z)$ as functions of height $z$) based on the readings of meteorological monitors measuring the temperature and surface wind speed at several levels (heights), including the temperature of the earth level, using the method of gradient observations.

**Table 1**. Averaged values of meteorological parameters [1].

| Season/time | $\bar{u}$, m/s | $\bar{k}$, m²/s | $\bar{b}$, m²/s² | $L$, m | $v_*$ |
|---|---|---|---|---|---|
| winter, 20:00 | 3,057 | 0,684 | 0,041 | 17 | 0,139 |
| spring, 20:00 | 3,836 | 11,743 | 1,129 | -5 | 0,232 |

$L$ – the scale of the surface layer of the atmosphere (the Monin-Obukhov scale).
$v_*$ – the dynamic velocity.

IOP Conf. Series: Materials Science and Engineering **487** (2019) 012009 doi:10.1088/1757-899X/487/1/012009

In order to obtain the simplest solution of the turbulent diffusion equation that describes the transport of the pollutant in the atmosphere in the framework of its surface layer, the functional dependences of these meteorological parameters were averaged over the height, and their mean values were used in the equation as constant coefficients [4, 5], given in table 1. Finally, the volumetric activity of the gas-aerosol radioactive pollutant propagating in the atmosphere were calculated in formula (1), in which the $\exp\left(-y^2/2\sigma_y^2(x)\right)/\sqrt{2\pi}\sigma_y(x)$, describes the Gaussian (transverse) dispersion, determined by formula (2), in which $\sigma_y(x)$ – is its dispersion; S (x, z) is the two-dimensional distribution function of the pollutant [2], determined by formula (3) [4, 5].

$$q(x,y,z)=\frac{S(x,z)}{\sqrt{2\pi}\sigma_y(x)}\exp\left(-y^2/2\sigma_y^2\right); \tag{1}$$

$$\sigma_y^2(x)=\bar{b}x^2/\bar{u}^2\left(1+ax\bar{b}/\bar{k}\bar{u}\right), \tag{2}$$

where $a = 0{,}015$;

$$S(x,z)=\frac{M}{2}\exp\left(-\left[\frac{\sigma_0 x}{\bar{u}}+\frac{w^2 x}{4\bar{k}\bar{u}}+\frac{w(z-h_{eff})}{2\bar{k}}\right]\right)\left\{\begin{aligned}&\frac{\exp\left(-\left[z+h_{eff}\right]^2\bar{u}/4\bar{k}x\right)+\exp\left(-\left[z-h_{eff}\right]^2\bar{u}/4\bar{k}x\right)}{\sqrt{\pi\bar{k}\bar{u}x}}-\\&-\frac{(2\beta-w)}{\bar{k}\bar{u}}\exp\left[-\frac{(2\beta-w)(z+h_{eff})}{2\bar{k}}+\left(\frac{2\beta-w}{2\bar{k}}\right)^2\frac{\bar{k}x}{\bar{u}}\right]\times\\&\times erfc\left[\left(\frac{2\beta-w}{2\bar{k}}\right)\sqrt{\bar{k}x/\bar{u}}+\frac{(z+h_{eff})}{2\sqrt{\bar{k}x/\bar{u}}}\right]\end{aligned}\right\}, \tag{3}$$

where $M$ – the value of the release rate (Bq /s); $w$ – is the gravitational velocity of deposition of the radioactive pollutant (m/s); $\sigma_0$ – is the constant of washing out of the pollutant from the atmosphere ($s^{-1}$); $\beta$ – is the rate of dry precipitation of the pollutant (m /s); $h_{eff}$ – is the emission effective height (m) (figure 1).

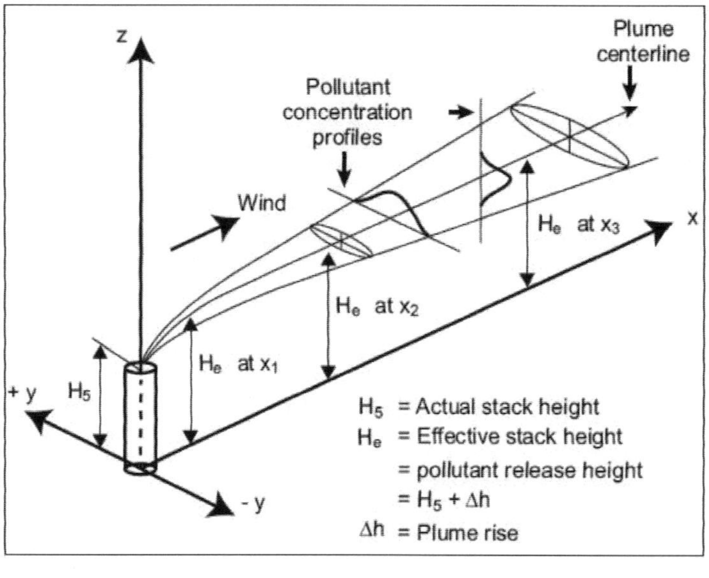


**Figure.1.** Schematic of the plume (Schulze and Turner, 1996).

## 2. Methods

### *2.1. Integral method*

The calculation of dose characteristics by the integral method [6], in which the volume of the radioactive pollutant that is propagating in the atmosphere is represented as a set of elementary sources with the corresponding coordinates, in each of which the pollutant distribution is assumed uniform. The distance $r$ between the detection point and the corresponding elementary source is determined by formula (4), and the value of the dose rate is calculated as an integral of the set of point sources by the formula (5).

$$r = \sqrt{(x - x_0)^2 + (y - y_0)^2 + (z - z_0)^2}\,, \tag{4}$$

where $x$, $y$, $z$ – coordinates of the point source; $x_0$, $y_0$, $z_0$ – coordinates of the detection point.

$$D'(x_0, y_0, z_0) = \frac{E \cdot \eta \cdot \mu_a}{4\pi} \cdot \int_V B(E, r).q(x, y, z) \cdot \frac{e^{-\mu \cdot r}}{r^2} dV, \tag{5}$$

where $E$ – energy of photons; $\eta$ - yield; $B(E, r)$ - build up factor [7]; $\mu_a$, $\mu$ – linear absorption coefficient and linear attenuation coefficient respectively; $V$ – volume of the source.

### *2.2. Monte Carlo Method*

At present, the basic methods for calculating the dose rates from radioactive pollutant of the underlying surface, which are used to estimate the dose rates from external radiation source, are based either on the concept of beam equilibrium [8] or on integral methods, taking into account the buildup factor [6]. It is advisable to estimate the dose rate for radioactive pollutant of air – within the framework of the concept of beam equilibrium – at distances from the emission source of more than 700 m, and at shorter distances – by an integral method [4]. In this case, the main error source that can be accrued is from distribution function of the radioactive pollutant determined by the formulas (1) – (3), since it contains a number of meteorological parameters measured with a certain error, and the error in calculating the buildup factor.

Monte-Carlo method (MC) is a numerical simulation of the physical process of photon interaction with matter based on the randomness of the photon scattering event, independent of its prehistory [9–12]. MC is free from the shortcomings typical of the integral method, but, unfortunately, it takes much more run time, to reduce the calculation error it needs more statistics, which is provided by a large number of stories, which consider random selection of elementary sources, forming volumetric radioactive pollutant. However, the technical capabilities of the current computer technology make it possible to use this method without any limitations [13–15].

## 3. Results

When calculating the dose rate of external irradiation from a volume source - plume of a radioactive pollutant - attention should be given for improving the distributions of this functional in the area of small (NPP site) distances from the source. The interest in this range of distances is due to the fact that a high gradient is observed in it in the distribution of the surface concentration of the radioactive pollutant, in connection with which the conditions of radiation equilibrium are violated [4] (see figure 2), which, in turn, can result to a significant error in estimating the dose rate. The greatest agreement of dose rate calculations is observed at distances from 0.7 to ~ 10 km from the source. At distances less than 0.7 km, the condition of the beam equilibrium is not fulfilled, and therefore the preference should be given to calculations by the Monte Carlo method and the integral method. Figure 3 shows the dose rate distributions from the volume source on its axis under the emission of $^{85}$Kr under a steady state of the atmosphere (winter), (curve 1) and unsteady state (spring), (curve 2). The calculations were carried out using the integral method. The region of the maximum on the curves

IOP Conf. Series: Materials Science and Engineering **487** (2019) 012009 doi:10.1088/1757-899X/487/1/012009

corresponds to the high density of the radioactive pollutant that appears in the atmosphere at small distances from the source. As the pollutant diffuses, the dose rate decreases, as curves 1 and 2 show. As for the shown nature of the distribution of dose rate for various states of atmospheric stability, this fact is explained by the significantly different distribution of the radioactive pollutant in the atmosphere, i.e. character of the distribution of the function q (x, y, z), determined by the expression (1), for various states of stability of the atmosphere. Figs. 4 and 5 show the results of calculating the dose rate for the steady (Figure 4) and unsteady (Figure 5) states of the atmosphere by the integral method (curves 1) and MC (curves 2) in these figures.

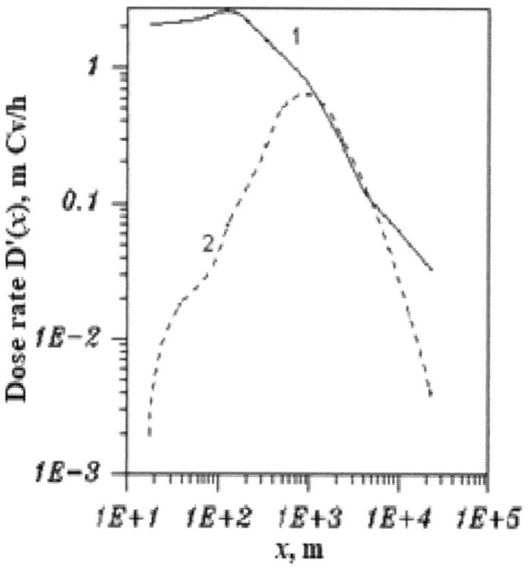


**Figure 2.** The dose rate distribution as a function of distance from the source during a hypothetical accident [4]: 1 – integral method; 2 – concept of radiation equilibrium.

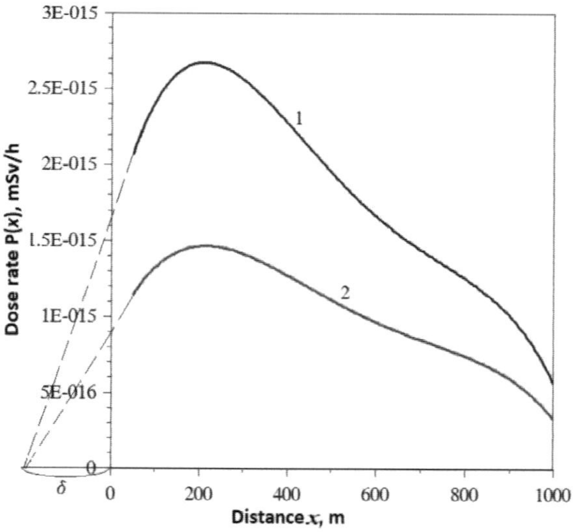


**Figure 3.** Dose rate distribution P (*x*) as a function of distance x from the source (during a hypothetical accident) using the integral method for steady (winter), curve (1) and non-steady (spring), curve (2) states of the atmosphere.

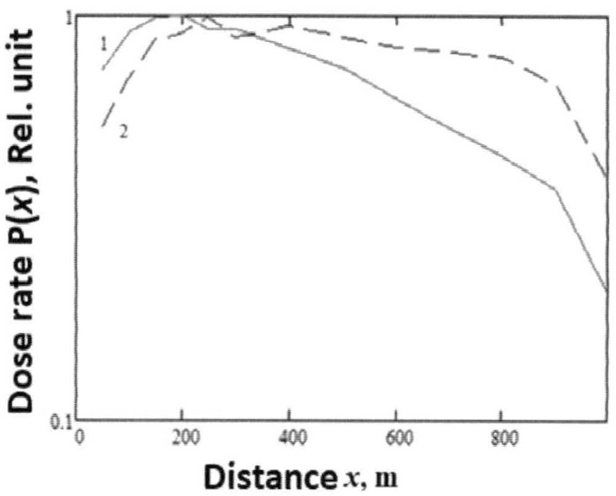


**Figure 4.** Dose rate distribution P (*x*) as a function of the distance *x* from the source (at a hypothetical accident) for the steady state of the atmosphere (winter) using the integral method, curve (1) and the Monte Carlo method, curve (2).

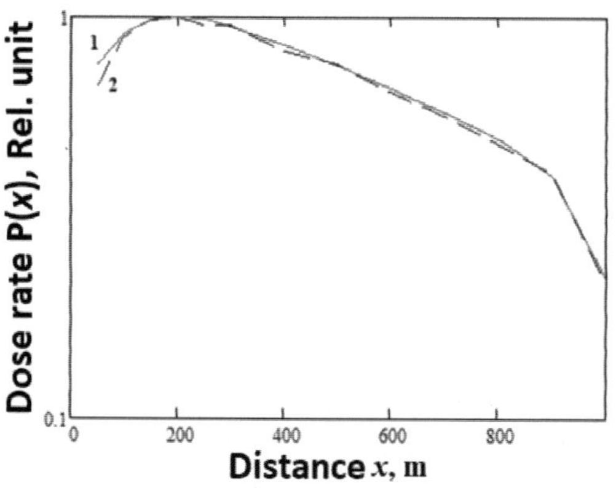


**Figure 5.** Dose rate distribution P (*x*) as a function of the distance *x* from the source (at a hypothetical accident) for the unsteady state of the atmosphere (spring) using the integral method, curve (1) and the Monte Carlo method, curve (2).

When forming the curves, there are two exclusive effects: the accumulation of volumetric activity in the atmosphere and its dispersion both due to the airflow and due to the transverse Gaussian broadening of the emission, as noted above.

In general, the results of calculations carried out by both methods agree satisfactorily, but much better with an increase in distances of ~ 700 m. The decrease in dose rate with increasing distances is due to a decrease in the volume activity of the radioactive pollutant that arises from its Gaussian broadening, which increases with increasing distance at source, and due to its decrease as a result of wind transfer. Scattering of the radioactive pollutant with increasing distance leads to its more even distribution in the atmosphere, and when calculating by the Monte Carlo method, in this case, a large number of histories are not required.

At such distances, an estimate of the dose rate can be obtained within the framework of the concept of beam equilibrium, as demonstrate Figure 2. On the contrary, at small distances from the source of emissions, the distribution density of volumetric activity is much higher than on large ones, and its volume is much less. As a result, in this area, much more histories are required to evaluate the dose rate than at large distances from the source. This is indicated by the fact that in both the steady and unsteady states of the atmosphere in the region of small $x$, for example, in the region of the maximum of the curves, the difference in the calculation results is 30-40%, and at large distances, the discrepancy is insignificant.

Comparing the results of calculations near the source, it should be noted that calculations of the dose rate in this region should indeed not be carried out, based on the concept of radiation equilibrium. One should also give attention to the value of the dose rate in the region of negative values of $x$ in Figure 3-5, i.e. in the region δ behind the vent tube. The absence of a radioactive pollutant in this region, as it blows the wind into the region of large $x$, only indicates that the inhalation dose in this region is zero, but the dose rate of external exposure, strictly speaking, in it will be nonzero at the expense of the scattered and non-scattered gamma radiation of a radioactive pollutant, since it is isotropic. An estimate of the dose rate values of the region δ can be different from zero, and it is easy to obtain by linear extrapolation of the dose rate values given in figure 3 with the corresponding state of atmospheric stability.

One of the main characteristics of radioactive emissions is the magnitude of the emission power $M$ ($Ci/s$), that is used in the solution of the turbulent diffusion equation (3). Indeed, this quantity plays the role of a normalization factor, since in all functions of the form (5) it will be simply a constant that is derived from the sign of the integral. It is enough to represent the volume activity $q(x, y, z)$ in (3) in the form of (1), in which the function $S(x, z)$ will be represented by formula (3). The significance of this parameter cannot be overestimated, given that all characteristics of radioactive pollution in the form of an air basin and the underlying surface or dose exposure on personnel and population will depend on the volume activity $q(x, y, z)$.

Given these features, an estimate of this value can be obtained by comparing the measured dose rate and its calculated value, determined by the formula (5). This can be done most simply if the detector is located in an area in which the distribution of volumetric activity can be considered uniform. Indeed, by measuring the value of the dose rate, in this case, with the coordinates ($x_0$, $y_0$, $z_0$), the number of which obeys the rule of the necessary number, and placing them on the territory of the site in a large distance from the center and a uniform distribution along the azimuth, for example, along the Archimedes spiral [1], the desired parameter $M$ can be found in the form of:

$$M = \frac{4\pi D'(x_0, y_0, z_0)}{E \cdot \eta(E) \cdot \mu_a(E) \cdot q_0 \cdot \int\limits_V B(E,r)\left[\exp(-\mu(E)r)/r^2\right]dV}, \tag{6}$$

where the distribution $q(x, y, z)$ can be considered uniformly distributed and equal to $q_0$. In this formula, in addition to $D'(x_0, y_0, z_0)$ the measured parameters are the energy of γ-photons $E$ and the volume activity $q_0$. When determining the parameter $M$, all other characteristics of radioactive pollutants of the environment and dose exposure on the population can be carried out at a one emission rate $M$.

Another possibility for estimating the volumetric activity $q_0$ (Ci/m$^3$) with its known radio nuclide composition is the direct use of the vent tube, from which the inert gases emission. In this case, when estimating the dose rate $D'$ from the cylindrical gaseous uniformly distributed source, which is the internal area of the vent tubes, and comparing it with the analogous calculated, calculated, for example, in the last-mentioned parameter, the desired parameter $q_0$ can be given as a ratio

$$q_0 = \frac{4\pi D'(x_0, y_0, z_0)}{E \cdot \eta(E) \cdot \mu_a(E) \int\limits_{V_C} \left[\exp(-\mu(E)R)/R^2\right] dV}, \tag{7}$$

where $R = \sqrt{\left(h_{eff} - h\right)^2 + r^2}$, $dV = 2\pi r dr dh$, volume $V_C$ is characterized by the range:

$0 \le h \le h_{\text{eff}}$; $0 \le r \le r_0$, $r_0$ – is the internal radius of the vent tube. In this case, the buildup factor, according to [4], can be equal to 1. If the discharge rate $G$ ($m^3/s$) of radioactive pollutant through the vent tube is known, (for example, it can be measured with the help of a flow and ionization chambers placed at the top of the vent tubes [16]), then $M$ can be found as the product $G \times q_0$ (Ci/s), and multiplying the product by the emission time $\tau_{\text{в}}$, the total activity of radionuclides released into the atmosphere can be found:

$$Q_0 = G \times q_0 \times \tau_{\text{в}}, \tag{8}$$

The advantage of the above methods for estimating the radioactive characteristics considered above is that they allow us to determine these parameters in the on-line Mode.

## References

[1] Alalem E A, Elokhin A P, Ksenofontov A I and Fedorov P I 2017 Meteorological characteristics of the planed nuclear power plant in Jordan *Global Nuclear Safety* **3** 19–34

[2] Laihtman D L 1970 *Physics of the Boundary Layer of the Atmosphere* (L.: Hydromet. Publisher) p 340

[3] Bobyleva M M 1970 *Calculation of Turbulence Characteristics in the Planetary Boundary Layer of the Atmosphere* (Leningrad Hydrometeorological Institute) vol 40 pp 64–73

[4] Elokhin A P 2014 *Methods and Instrumentations of Radiation Monitoring Systems of the Environment* (Moscow: NRNU MEPhI) p 520

[5] Elokhin A P and Starodubtcev I A 2017 On the ecological situation at the territories adjacent to chemical and metallurgical facilities *Environ. Quality Management* **26** 23–43

[6] 1971 *Meteorology and Atomic Energy* Transl. from English. N L Byzova and K P Makhonko (L .: Gidrometeoizdat) p 618

[7] Mashkovich V P and Kudryavtseva A V 1995 *Shielding from Ionizing Radiation Manual* (M.: Energoatomizdat) p 496

[8] Gusev N G and Belyaev V A 1986 *Radioactive Dispersion in the Biosphere* (M: Energoatomizdat) p 224

[9] 1967 *Monte Carlo Method in the Problem of Radiation Transfer* Ed. Marchuk (Moscow: Atomizdat) p 256

[10] Leimdorfer M 1964 On the use of Monte-Carlo methods for solving gamma radiation transport problems *Nukleonik* **6** 14

[11] Zolotukhin V G, Kimel L R and Ksenofontov A I 1974 *The Radiation Field of a Point Mono-directed Source of Gamma Quanta* (Moscow: Atomizdat) p 160

[12] Sobolev I M 1973 *Numerical Monte Carlo Methods* (M .: Nauka) p 311

[13] Elokhin A P, Khmylov A N and Zhilina M V 2007 Method for assessing the consequences of radiation accidents in the premises of a reactor unit at a nuclear power plant with a VVER-

1000 reactor *Atomic Energy* **102** 254–62
[14] Zhilina M V 2010 Application of the Monte Carlo method in problems of radiation monitoring of the environment *Ecological Systems, Devices* **10** 3–12
[15] Elokhin A P, Zhilina M V and Parkhoma P A 2009 Features of scanning the underlying surface with an unmanned dosimetry complex *Atomic Energy* **107** 103–12
[16] Elokhin A P 2012 *Automated Systems for Monitoring the Radiation Environment of the Environment Tutorial* (M .: NRNU MEPhI) p 316

# Assessment of radiation environmental risk for the terrestrial ecosystem

**G V Lavrentyeva**

Bauman Moscow State Technical University (Kaluga Branch), Kaluga, Russia

E-mail: Lavrentyeva_G@list.ru

**Abstract**. In this work there is a detailed methodology for the assessment of the radiological ecological risk at the ecosystem level based on critical loads. Risk assessment consists of the following stages: hazard identification, identification of reference species and indicators, determination and analysis of critical loads on the basis of the dose-effect relationship, environmental risk assessment and analysis of risk functions. The experimental site is the territory, which has been exposed to radioactive contamination as a result of depressurization of the storage capacity of radioactive waste. On the basis of long-term (2010 – 2015) monitoring works, the input parameters for risk assessment were determined: the recipient of the impact is the terrestrial ecosystem; in the study of chemical and radioactive contamination of the territory, an environmental hazard factor was identified – Sr-90; the reference species - the mollusk *Fruticola fruticum* and reference indicators – the coefficient of accumulation of the Sr-90 of shells from the nettle and catalase enzymatic activity of soils. It was determined the acceptability of radiation risk for the terrestrial ecosystem on the basis of critical loads on the catalase activity of soils and unacceptability when accounting the coefficient of accumulation of Sr-90.

## 1. Introduction

ICRP in its publication [1] points out the need to provide direct evidence of the protection of biota. It reflects the eco-centric principle of rationing of the radiation factor. In this regard, many recent works are devoted to the development of ecological approaches to the rationing of radiation effects on ecosystems. It should be noted that there are two approaches to the regulation of environmental pollution. On the one hand, it is possible to ration the content of pollutants in environmental objects, on the other hand, the degree of transformation of the environment as a result of its contamination. One of the indicators of sustainability of ecosystem to anthropogenic impact, including radiation, is the magnitude of critical loads. Critical load is an indicator of the sensitivity of the ecosystem, which determines the maximum acceptable of contamination. The environmental risk is the final stage in estimating critical loads for ecosystems.

The aim of this work is assessment of radiation environmental risk on the basis of critical loads for the terrestrial ecosystem in the zone of influence of the radioactive waste storage facility.

The research area is the territory, which was exposed to radioactive contamination in 1998 as a result of depressurization of the storage capacity of radioactive waste [2].

## 2. Materials and methods

Soil, nettle nettle (*Urtíca dióica*) and terrestrial mollusk *Fruticicola fruticum* samplings were done to determine the input parameters for risk assessment. Samples were selected at 42 local controlled sites. The soil was selected by means of a specialized sampler with a set of tame soil drills by Edelman ("Eijkelkamp", The Netherlands). Samples of mollusks were sampled from plants and from soil under plants in the amount of 25-30 individuals from one local site. From the local site, the whole nettle biomass was selected.

Measurement of specific activity with preliminary radiochemical isolation of Sr-90 in samples of soil, plant, terrestrial mollusks was carried out on the scintillation β-spectrometer BETA 01C. The determination of Cs-137 in the samples was carried out by ɣ-spectrometry on the ɣ-spectrometer Accuspec, Canberra Industries. The concentration of heavy metals was determined by the method of atomic emission with inductively coupled plasma (ICP AES Varian Liberty II). The specific activity of Sr-90, Cs-137 and the concentration of heavy metals in each sample were determined in triplicate.

The activity of catalase activity was carried out according to the amount of molecular oxygen released per gram of soil per minute.

The statistical processing of the experimental data was carried out using the program R.

## 3. Results and discussion

The methodology for environmental risk assessment consists of the following stages: hazard identification, identification of reference species and indicators, determination and analysis of critical loads on the basis of the dose-effect relationship, environmental risk assessment and analysis of risk functions.

### 3.1. *Hazard identification*

This stage of risk assessment is aimed at determining the recipient ecosystem, the impact indicators on the ecosystem and the assessment of their values. The aquatic and terrestrial ecosystem were considered as possible recipients. The chemical pollution of the ecosystems (K, Na, Ca, Mg, Sr, Al, Fe, Mn, Zn, Ni, Cu, Cr, Co, Cd, Pb) and radioactive contamination (Sr-90, Cs-137) were selected as impact indicators. It has been established that chemical pollution cannot be considered as indicator of impact on ecosystems, because there is no excess of standards.

Analysis of data from 1998 to 2014 of radioecological monitoring of contamination of Cs-137 and Sr-90 of water objects shows that the aquatic ecosystem cannot be an exposure recipient [3]. It is a local contamination of soil with the radionuclide Cs-137 and the radionuclide is mainly (50-53%) in a fixed form of location. Therefore, Cs-137 cannot be an impact indicator in assessing environmental risk. However, radioactive contamination of soil by Sr-90 was detected throughout the area with its specific activity from a control value of 19.7±11.1 to 5203±89 Bq/kg. The radionuclide is in the soil at 69-83% in its exchange form. Thus, the terrestrial ecosystem is the recipient of the impact and Sr-90 is the impact factor. The area of exposure is determined to be 0.54 hectares.

### 3.2. *Identification of reference species and indicators*

To assess the radiation environmental risk, we propose to use the term referent indicator – radiation-induced effect in biota representatives at the level of an individual, species, population or ecosystem as a whole, which can be described by a reliable model having a threshold value. The threshold value will act as a critical load in the risk assessment.

The terrestrial mollusk Fruticicola fruticum is justified as a reference species, the reference indicators are the coefficient of accumulation of the Sr-90 of shells from the nettle and catalase enzymatic activity of soils. It should be noted that mollusks have long been recognized as a convenient bioindication for environmental pollution due to high accumulation rates of heavy metals and radionuclides, widespread prevalence, ease of identification, short life cycle [4, 5]. In this case, shells of terrestrial mollusks contain about 95-99% calcium carbonate, so the shell of the mollusk is a high accumulation of strontium – as a chemical analogue of calcium.

It is known that radioactive contamination of soil can disrupt its homeostasis and lead to degradation. It causes a change in the functional and biochemical activity of soil microorganisms. At the same time, many years of research have established a high efficiency of environmental soil diagnostics through biochemical methods, among which the determination of enzyme activity is the most promising.

*3.3. Determination and analysis of critical loads on the basis of the dose-effect relationship*
The input parameters for the determination of critical loads are the specific activity of Sr-90 in soil (19.7±11.1 – 5203±89 Bq/kg), nettle (22.3±13.4 – 10596±195 Bq/kg), shells of mollusks (76±11 – 17640±2646 Bq/kg).

It was found that the change in catalase activity can be described by a reliable model with a threshold value of 21.3±5.9 $cm^3O_2/g*min$ [6]. The input parameter for calculating the risk is the critical load of 1858 Bq/kg of Sr-90 in the soil (figure 1).

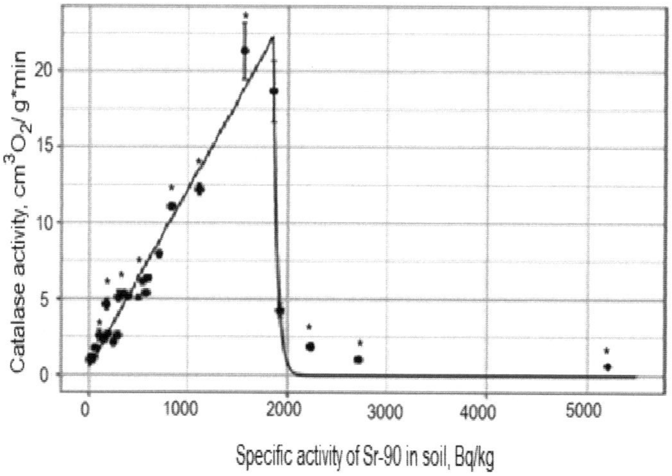


**Figure 1.** Change in catalase activity with increasing Sr-90 content in soil.

The analysis of the coefficient of accumulation of Sr-90 by shells of mollusks makes it possible to reveal a significant change in the indicator in the range of variation of the specific activity of Sr-90 in the nettles from 22 to 10.5 kBq/kg (figure 2). The decrease of the indicator is described by a power law dependence. The specific radionuclide activity in the nettle 2138 Bq/kg is the critical load.

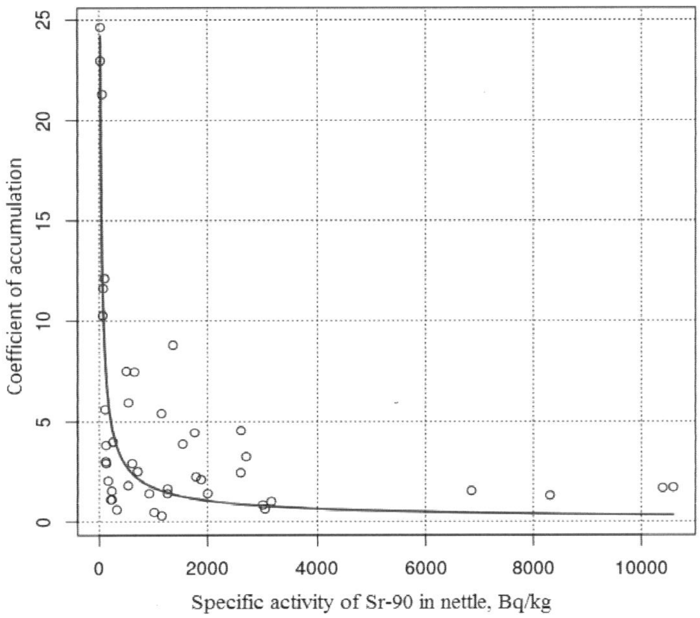


**Figure 2.** Change in accumulation coefficient.

*3.4. Environmental risk assessment and analysis of risk functions*

The environmental risk assessment was carried out through GIS-technologies. An acceptable value for the assessment of ecosystem risk is the 95% protection of the ecosystem: an acceptable risk is considered to be that the area with exceedance of critical loads does not exceed 5% of the total area [7].

The excess of the critical loads was calculated as the difference between the specific activity of Sr-90 in the soil (nettle) of the local area and the critical load. After calculating exceedances of critical loads for each local area, program R was used to determine the proportion of the area with excess of critical loads.

Analysis of calculation of exceedance of critical loads shows that the area characterized by an exceedance of the critical load for catalase activity is 0.21%, which does not exceed an acceptable value of 5% (figure 3). Environmental risk is characterized as acceptable.

The analysis the corresponding experimental data for coefficient of accumulation of Sr-90 showed an excess of an acceptable value and the share of the area exceeding the critical load was 5.75% (figure 4). Thus, the risk of the storage impact on the terrestrial ecosystem should be characterized as unacceptable.

The values of areas with exceedance of critical loads allow to proceed to determine the probability of development of negative changes in the ecosystem on the basis of analysis of risk functions. This analysis shows that the risk of exceeding the critical load taking into account the catalase activity of the soil is approaching 0, and taking into account the coefficient of accumulation Sr-90 exceeds 5%, which confirms the risk assessment through GIS mapping.

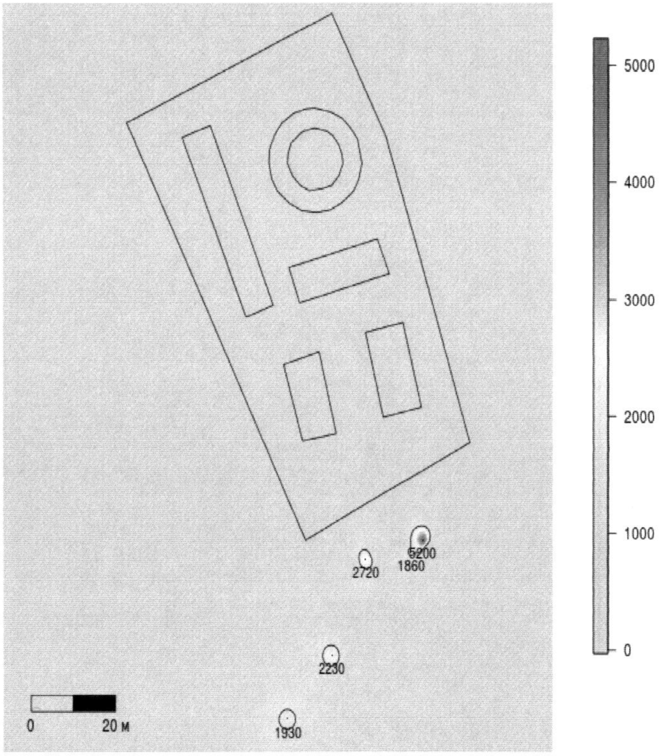


**Figure 3**. GIS-map for critical loads of catalase soil activity.

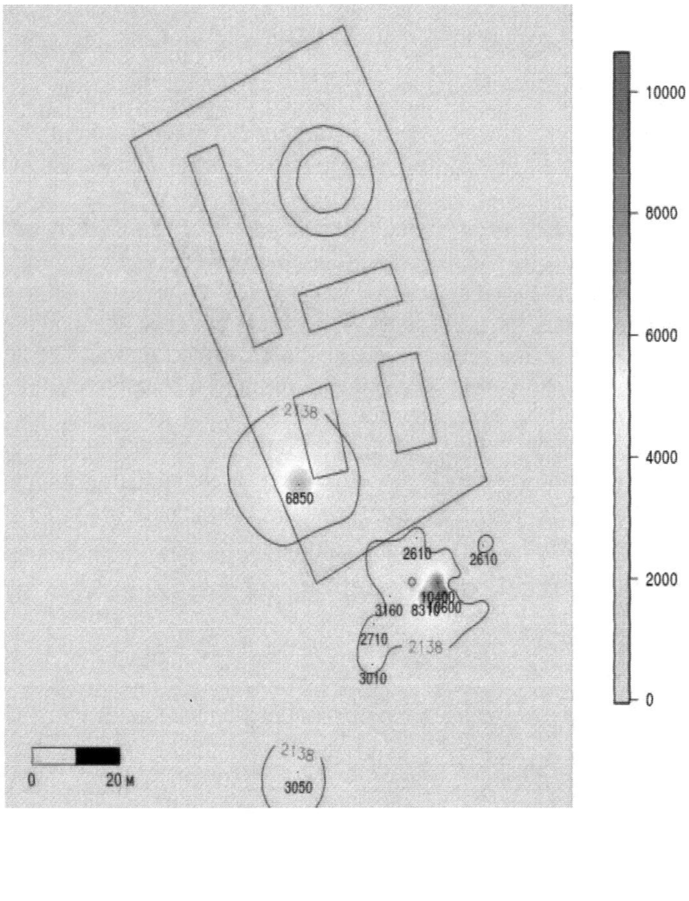


**Figure 4**. GIS map for critical loads on the factor of accumulation Sr-90

## 4. Conclusions

In this paper, it was assessed the radiation environmental risk for the terrestrial ecosystem and was considered the methodology of risk assessment based on critical loads. On the basis of an experiment in natural conditions it was determined the acceptability of radiation risk for the terrestrial ecosystem on the basis of critical loads on the catalase activity of soils. However, the risk is unacceptable when it is evaluated on the basis of an analysis of the coefficient of accumulation. It should be noted that the unacceptability of the risk in this case is characterized by exceeding the value of 5% by 0.75% from the possible 95%. It approximates the acceptability of risk, if we take into consideration uncertainty in risk assessment.

With considering the assessment of radiation environmental risk the radioactive waste storage facility can be considered an environmentally safe, even if reservoir depressurization is taken into account.

## References

[1] 2007 ICRP Publication 103. The 2007 Recommendations of the International Commission on Radiological Protection. *Ann ICRP* p 332
[2] Lavrentyeva G V 2014 *J. Environ. Radioact.* **135** 128–34
[3] Lavrentyeva G V, Silin I I, Kozmin G V, Vasileva A N, Synzynys B I, Glushkov Ju M and Momot O A 2012 *Water: chemistry and ecology* **12** 26–31
[4] Majone F, Brunetti R, Gola I and Levis A G 1987 *Mutat.Res.* **191** 157–1
[5] Golubev A, Sikorski V and Stoliar O. 2011 *Radioprotection* **46** 393–9
[6] Lavrentyeva G V, Zaharova V R, Mirzeabasov O A and Synzynys B I 2017 *XIII Int. Youth Scientific and Practical Conf. «Future of atomic energy – AtomFuture 2017»* pp 137–42
[7] Bashkin V N and Priputa I V 2011 *Problems of Risk Analysis* **8** 8–21

IOP Conf. Series: Materials Science and Engineering **487** (2019) 012011 doi:10.1088/1757-899X/487/1/012011

# Modelling of distribution of aerosol emission with account of buildings and structures of NPPs

**M Mehdi[1,2], M P Panin[1] and B Mohammedi[2]**

[1]National Research Nuclear University MEPhI (Moscow Engineering Physics Institute), Moscow, Russia
[2]Nuclear Research Center of Birine, Algeria

E-mail : mehdiyahia1@hotmail.fr

**Abstract**. In the lack of experimental data and due to the rapid development of numerical modelling and computer hardware, computational fluid dynamics (CFD) is widely used to study wind field and contaminant transport around buildings. The ability to predict quantitative values for these processes is not so clear. CFD simulations work under neutral atmospheric conditions and are validated by the Gauss model. The Gaussian model is a mathematical distribution of the concentration of contaminants from NPP ventilation pipes. It is the most recommended model by the IAEA and most widely used because 1) it gives results consistent with experimental data, 2) it is quite easy to use, and 3) it is consistent with the aleatory nature of turbulence. Reynolds-averaged Navier-Stokes equations is the most widely used method in turbulent flow modeling and simulation. In this paper, the prediction of flow accuracy and propagation around buildings with a stack was examined using $k$-ε model. The numerical results were compared to Gauss model results. Contaminant dispersion was well predicted with $k$-ε model. It was also confirmed that the concentrations predicted by the CFD models were more diffused than those of the Gauss model and that the results depend on the position of the obstacles.

## 1. Introduction

The transport of atmospheric pollutants in the environment is influenced by many complex factors. Among them, are wind conditions and urban morphology which are the most relevant to pollutant dispersion. Examples of wind conditions are wind speed and turbulence intensity. The higher the wind speed, the more mixing of pollutants with fresh air and the greater is the dilution - in other words, the lower is the concentration of pollutants detected in the wind flow. As for urban morphology, the more complex it is, the more it increases vortex structures in the wake of buildings. These recirculation zones tend to trap pollutants and increase local concentrations, which presents a hazard when the fresh air intakes of a building are located in a contaminated area, as soon as the risk of ingesting pollutants is increased [1–3].

This episodic phenomenon is present in all high-density urban areas; however, the literature offers little information and recommendations to avoid the problems that arise from it. In order to limit indoor air contamination by ingestion of pollutants from the outside, a better understanding of pollution aerodynamics and mechanisms of transport is necessary.

The research methodology used employs computational fluid dynamics (CFD) to study the dispersion of pollutants around a group of buildings. This tool provides detailed information on flow profiles and concentration (or dilution) fields by solving the flow equations throughout the calculation range. The reliability of numerical simulations is one of the main concerns of this study. CFD is based on the analysis of fluid flows, mass transfer and related phenomena, solving a subset of Navier Stokes' conventional equations at finite locations on a grid. It simultaneously produces results relating to flow characteristics at all points in space. Based on fundamental physics, CFD is used in a wide range of basic research and practical applications. In the field of urban wind studies, CFD has proven to be a promising technology because of its flexibility in modeling complex geometries, such as those of cities with high building densities. CFD is not inherently limited by similarity constraints (as in wind tunnels), so it should be possible to digitally simulate all aspects of pollutant dispersion and its interactions with the environment [4]. That said, while CFD offers some advantages, it requires special care to provide reliable results. A number of parameters, such as grid size, discretization scheme, choice of turbulence model and boundary conditions, need to be controlled and validated by systematic comparisons with experimental data or other highly accurate methods [5].

One of the first studies on the complexity of the flow field around a body (representing a single building) and the relative effectiveness of different models of turbulence was carried out by Murakami and Mochida [6]. In this study, the distribution of velocities obtained from three-dimensional simulations of continuous flow around a cubic model was compared with wind tunnel results to evaluate the accuracy of the standard k-ε turbulence model. An examination of the distribution of kinetic energy (k) revealed that the production level of k at the windward angle of the model was largely overestimated. Among other subsequent research, the study suggested changes to turbulence modeling in the k-ε model.

Meroney et al [7] studied the flow field and dispersion around several shapes of buildings. They compared the results obtained in the wind tunnel to the measurements obtained in the wind tunnel. The purpose of these comparisons was to determine whether commercial software could be used to correctly simulate engineering wind problems. The authors found that numerical simulations invariably overestimated the surface concentrations downwind of the emission source. They attributed the observed differences to the inability of the Reynolds numerical model, on average, to reproduce the intermittency of the flow in the recirculation zones visualized in the wind tunnel. Moreover, although the concentration profiles were well reproduced, the amplitudes were frequently higher by an order of magnitude than those measured in the wind tunnel. With regard to the profiles of pressure, numerical predictions have been shown to be reasonably accurate, and amplitudes sufficiently satisfactory to allow engineering calculations, implying that mean pressure fields are less sensitive than other criteria to features of the digital model. The study finally revealed that the turbulence model RSM produced somewhat more realistic results than the SKE and RNG models.

## 2. Gaussian Model

Is one of the most widely used models to represent numerically the movement and dispersion of effluent from an emission point. Figure 1 illustrates in a simplified way a Gaussian model representing the contaminants emitted by a pipe.

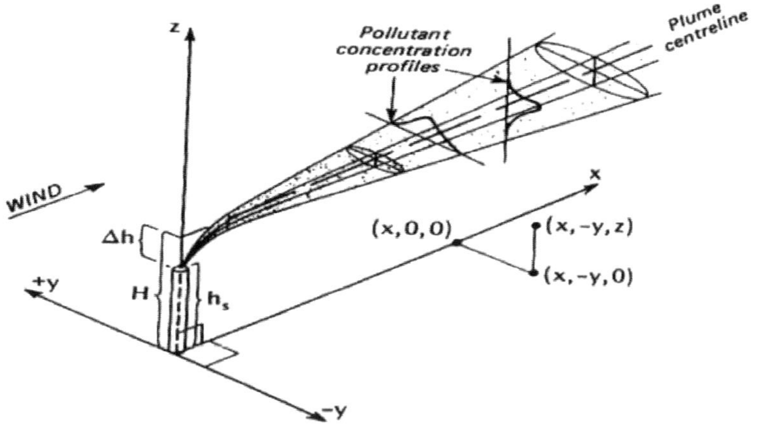


**Figure 1.** Plume dispersion coordinate system.

Gaussian dispersion model

$$C(x,y,z;H) = \frac{Q}{2\pi\sigma_y\sigma_z u} * exp\left[-\frac{1}{2}\left(\frac{y}{\sigma_y}\right)^2\right] * \left[exp\left[-\frac{1}{2}\left(\frac{z-H}{\sigma_z}\right)^2\right] + exp\left[-\frac{1}{2}\left(\frac{z+H}{\sigma_z}\right)^2\right]\right]$$

where C (x, y, z): average contaminant concentration at one point in the space in $Bq/m^3$;
Q: radioactivity flow rate emitted by the source in Bq/s;
u: average horizontal wind speed in m/s at the height of emission;
$\sigma_y$ et $\sigma_z$ standard deviations of the horizontal and vertical Gaussian distribution in meters;
H: total plume elevation in meters which is equal to h + Δh: actual pipe height plus the immediate plume height.

The difficulty in a Gaussian model is the evaluation of turbulent diffusion. These are the standard deviations of distribution $\sigma$ which represent the turbulent diffusion along the y and z directions in this equation. These coefficients depend on flow characteristics, turbulence, atmospheric stability and surface roughness. They are experimentally determined from large-scale trials. There are several correlations for determining standard deviations including those of Briggs [8], which depend on distance to source and atmospheric stability classes and those of Doury [9], which depend on transfer time.

## 3. CFD simulations

ANSYS-FLUENT 14.5 [10], has being used for simulation**.** The purpose of numerical predictions is to solve all partial differential equations applicable to the flow of any fluid, including the wind flow in the atmosphere. These equations are based on the fundamental laws of mass conservation and amount of motion (Navier-Stokes equations) in the computational domain.

### *3.1 The Computational domain*

In this study the atmospheric dispersion is modeled around the Rostov nuclear station. The Rostov nuclear station consists of four blocks, the ventilation pipe is placed in the middle of the Y axis and 320 m from the X axis with a height of 100 m. The dispersion simulation area is 1520 m x 800 m x 300 m. For the dimensions and definition of the location of buildings and structures, we used satellite maps obtained from the Yandex site.

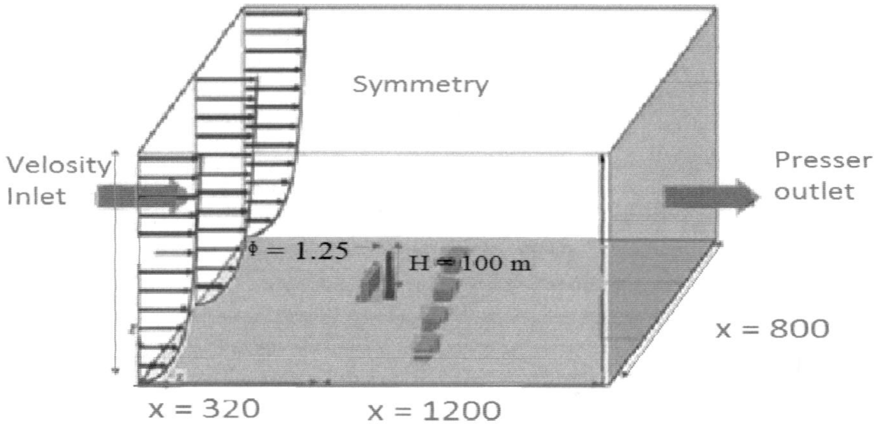


**Figure 2.** Schematic sketch of geometry and boundary conditions.

*3.2 The Boundary Conditions*

For the k-ε model, Durbin and Petterson Reif [11], proposed vertical profiles for mean wind speed *u*, turbulent kinetic energy k and turbulence dissipation rate ε.

$$u(z) = \frac{u^*}{K}\left[ln\left(\frac{z}{z_0}\right)\right]$$

$$k = \frac{u^{*2}}{\sqrt{C_\mu}}$$

$$\varepsilon = \frac{u^{*3}}{Kz},$$

where z is the height co-ordinate, $u^*$ the friction velocity, $K$ the von Karman constant (0.40-0.42) and $C_\mu$ a model constant of the standard $k$- $\varepsilon$ model.

## 4. Results and discussion

At the entrance of the simulation domain, the wind speed profile, in red on figure 3, gives a minimum speed of 0 m/s on the ground at z = 0 m and a maximum speed of 5.7 m/s at z = 100 m (the height of the stack). For the position between the blocks, at x = 270 m, the speed profile from z = 0 m to z = 60 m (the height of the buildings) the profile becomes complex due to the eddies created by buildings. Far away from buildings (x = 1200 m) the wind speed profile returns to ordinary shape.

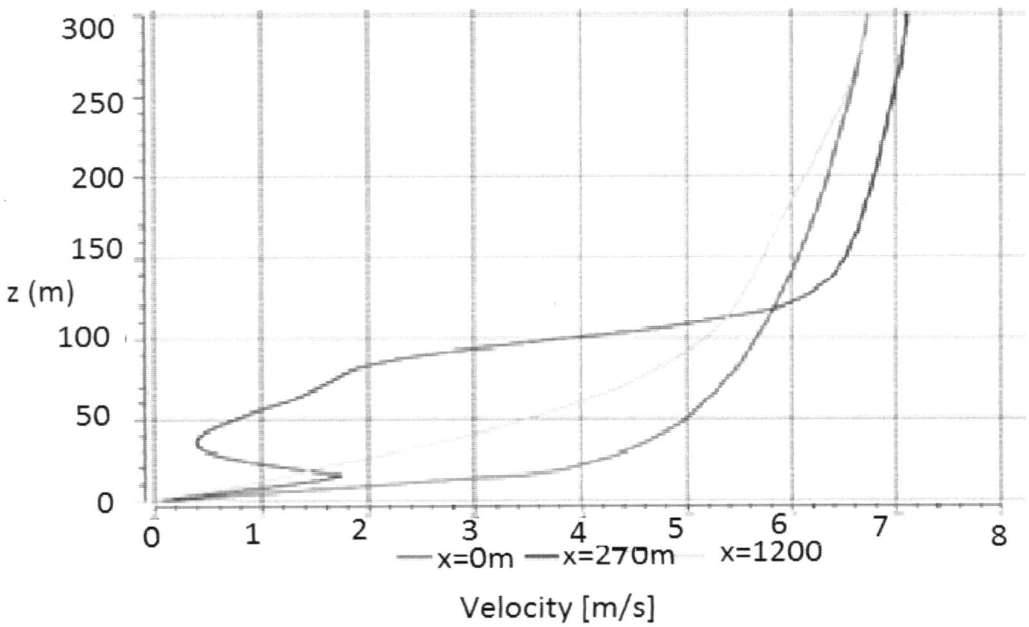


**Figure 3.** Simulated horizontal velocities at three locations.

IOP Conf. Series: Materials Science and Engineering **487** (2019) 012011 doi:10.1088/1757-899X/487/1/012011

A large recirculating vortex can be seen over the entire height of the stacks and blocks, at low speed (figure 4).

As expected and quite naturally the blocks influence the wind path around the site (figure 5). Buildings in their wake create recirculations (eddies) slowing the flow. The blocks downstream of the flow have an influence that can be noticed by the bleu vectors (low speeds).

**Figure 4.** Velocity vectors (vertical cut).

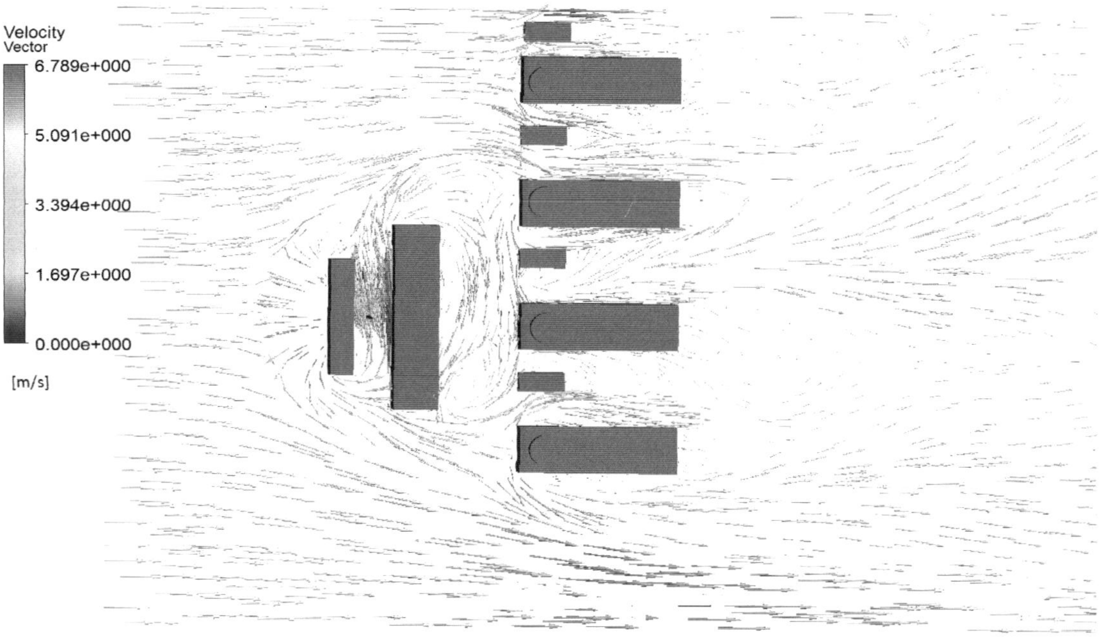


**Figure 5.** Velocity vector (top view).

Reduced interval between blocks creates a flow vortex which partially drives the flows (figure 6), contributing to an increase in their dilution downstream.

The blocks could have a "positive" effect on contaminant plume dilution. Indeed, the plume expanded vertically under the effect of the zone of turbulence created by blocks (figure 7). The plume is in a way less concentrated because it is more vertically extended.

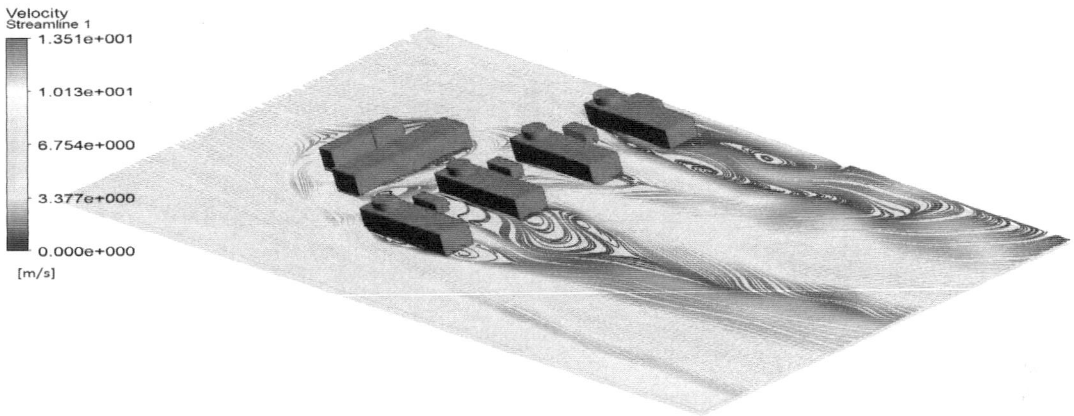


**Figure 6.** Streamlines velocity.

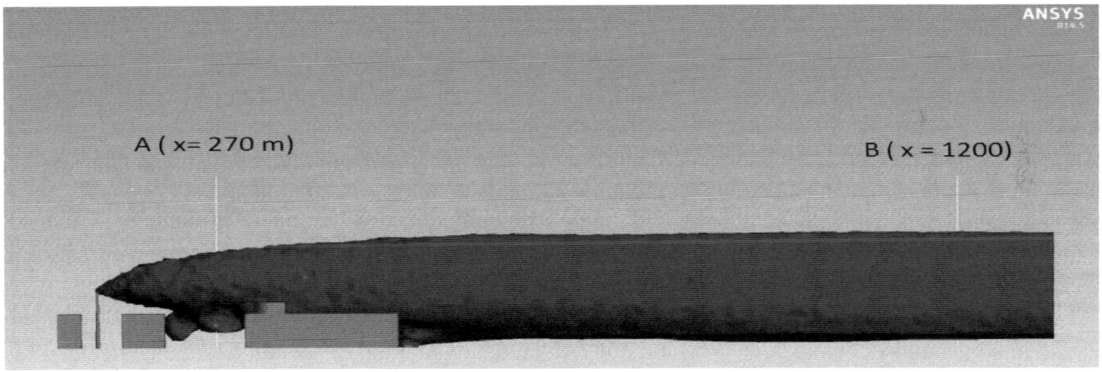


**Figure 7.** Isosurface of contaminant concentration.

The CFD (Ansys-Fluent) calculations were compared with the results of the Gaussian model (Briggs model) on a flat surface, the calculations are carried out in two lines parallel to the buildings and taken at the center of the plume y = 0 m, with x = 270 m between the buildings and x = 1200 a little far from the buildings, and z = 300 for the two lines, they show good consistency (figure 8). In this case, spatial dependence of the parameters of Gaussian distribution was taken from the models of the first generation of Briggs. When simulating the flow through buildings exits, the effect of the significant increase in the concentration from leeward side of the building (figure 8) was found due to the turbulence created by the flow around the building.

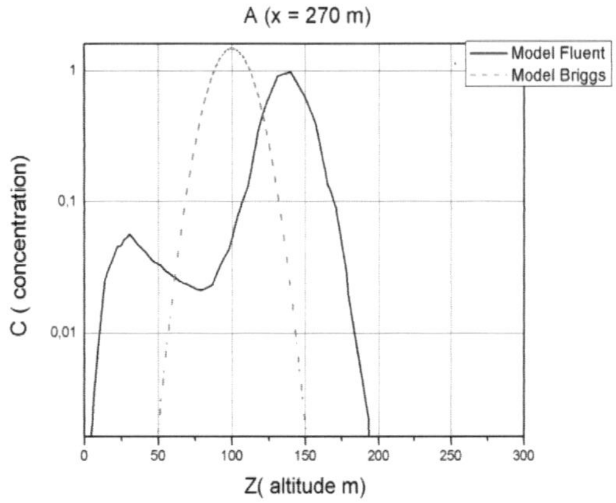


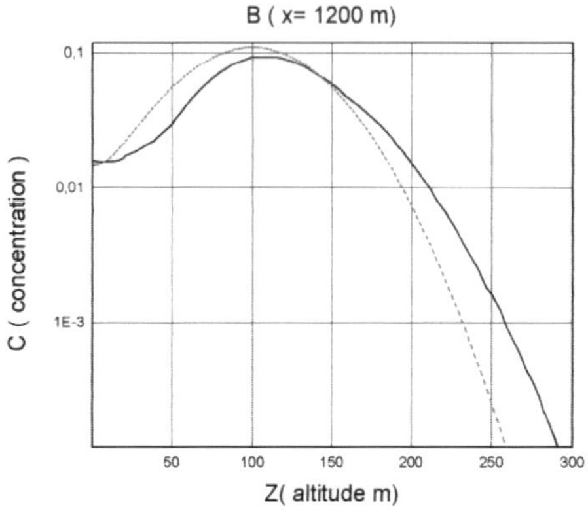


**Figure 8**. The vertical concentration profiles in lines A and B are indicated in Figure 7.

## 5. Conclusion

In this work a numerical study of the evolution of radioactive contaminant from an NPP ventilation pipe in the presence of obstacles is presented. This type of study allows the determination of areas where wind speed is accelerated by the presence of surrounding obstacles. The results essentially show that an obstacle in the path of a gas release can divert it either off course or/and from its direction; laterally or vertically. Near buildings, the concentration of radioactive emissions calculated by the Gauss model has a significant error.

## References

[1] Snyder W 1981 *Guidelines for Fluid Modelling of Atmospheric Diffusion*. EPA office of Air quality, planning and standards, Research triangle park, NC, EPA-600/8-81-009

[2] Schulman L L and Scire J S 1993 Building downwash screening modeling for the downwind recirculation cavity. *J. Air. Waste. Manag. Assoc.* **43** 1122–27

[3] Saathoff P, Gupta A, Stathopoulos T and Lazure L 2009 Contamination of fresh air intakes due to downwash from a rooftop structure *J. Air. Waste. Manag. Assoc.* **59** 343–53

[4] Merone R N 2004 Wind tunnel and numerical simulation of pollution dispersion: a hybrid approach. Working paper, Croucher Advanced Study Institute on Wind Tunnel Modeling, (Hong Kong University of Science and Technology)

[5] Blocken B, Stathopoulos T, Saathoff P and Wang X 2008 Numerical evaluation of pollutant dispersion in the built environment: comparisons between models and experiments *J. Wind. Eng. Ind. Aerod.* **96** 1817-31

[6] Murakami S and Mochida A 1988 3-D numerical simulation of airflow around a cubic model by means of the k-ε model *J. Wind. Eng. Ind. Aerod.* **31** 283–303

[7] Meroney R N, Leitl B M, Rafailidis S and Schatzmann M 1999. Wind-tunnel and numerical modeling of flow and dispersion about several building shapes *J. Wind. Eng. Ind. Aerod.* **81** 333-45

[8] Hanna S R, Briggs GA and Hosker RP 1982 *Handbook on Atmospheric Diffusion*. DOE/TIC11223 (U. S. Department of Energy, Oak Ridge, TN)

[9] Leroy C. 2010 A study of the atmospheric dispersion of a high release of krypton-85 above a complex coastal terrain, comparison with the predictions of Gaussian models (Briggs, Doury, ADMS4) *J. Environ. Radioact.* **101** 937–44

[10] ANSYS FLUENT User's Guide.

[11] Durbin PA and Petterson Reif BA 2001. *Statistical Theory and Modelling for Turbulent Flows* (John Wiley &Sons, Chichester, UK)

# Assessment of the radiation dose rate for a terrestrial mollusk during chronic Sr-90 irradiation

**G V Lavrentyeva[1], A E Oorzhak[2], R R Shoshina[2], A A Oudalova[2,3] and Yu A Kurachenko[3]**

[1] Bauman Moscow State Technical University (Kaluga Branch), Kaluga, Russia
[2] Obninsk Institute for Nuclear Power Engineering, National Research Nuclear University "MEPhI", Obninsk, Russia
[3] Russian Institute of Radiology and Agroecology, Obninsk, Russia

E-mail: Lavrentyeva_G@list.ru

**Abstract.** Radiation dose rate for a terrestrial mollusk *Fruticicola fruticum* was assessed under chronic Sr-90 irradiation in field conditions by three methods. The study-site is the territory nearby the radioactive waste storage facility in the Kaluga region. Due to a loss of integrity in one of the storage capacities, there was a leakage of radionuclides in 1998. Specific activities of Sr-90 in soil, plant and shells of terrestrial mollusk used as the input parameters for calculating the radiation dose rate are: 20±3 – 5203±785 Bq/kg; 22.3±13.4 – 10596 ± 195 Bq/kg, 76±11 – 17640±2646 Bq/kg, respectively. As a result of the calculations, the following radiation dose rates were obtained: from 10±2 to 1634±245 mGy/year with the J. Levinger and L.D. Marinelli formula, from 0.012±0.001 to 2.1±0.2 mGy/year with the Erica Assessment Tool, from 0.32±0.07 to 76±9 mGy/year by the Monte Carlo method. The Monte Carlo method gives the most adequate estimates, because the application of this method allows taking into account the maximum possible scenarios of mollusk irradiation, the maximum number of input parameters, including the weight and geometry of the organism, the dosimetric and physical features of the source of irradiation.

## 1. Introduction

In spite of the wide experimental and theoretical knowledge, methods for estimating radiation doses for biota are still under development. At the same time, many researchers work with the corresponding software packages. For example, the Erica Assessment Tool is most often used when assessing dose loads for reference animals and plants [1–5]. An application of the Erica Assessment Tool is recommended by the international organizations on the environment radiation protection. This methodology is developed in accordance with ICRP Publication 108 [6]; however, in view of the diversity of flora and fauna representatives, the possibility to estimate doses and dose rates by other methods is also not denied.

To calculate radiation doses, special mathematical models are also developed taking into account specific radioecological conditions and morphological features of an organism, which in some cases cannot be estimated with the Erica Assessment Tool [7–9].

In this paper, a comparative analysis is done for three approaches to calculate an absorbed radiation dose rate on an example of a terrestrial mollusk *Fruticicola fruticum*: the Erica Assessment Tool, a

IOP Conf. Series: Materials Science and Engineering **487** (2019) 012012 doi:10.1088/1757-899X/487/1/012012

classical computational method by the J. Levinger, L.D. Marinelli formulas, and the Monte Carlo method.

As study site in this work, the territory nearby the radioactive waste storage facility in the Kaluga region of Russia is considered. In 1998, there was a leakage of radionuclides from one of the storage capacities due to a loss of its integrity [10]. From the long period of observations (1998 – 2015), it has been well established that Sr-90 is the main radionuclide responsible for the radioecological situation in this study site [10].

## 2. Materials and methods

Samples of the soil, terrestrial mollusks *Fruticicola fruticum* and nettle *Urtíca dióica* (as the main food for mollusks) were taken from 42 local sites within the surveyed territory nearby the radioactive waste storage facility. Soil was selected by means of a specialized Edelman sampler ("Eijkelkamp", The Netherlands). Mollusks were collected from plants and from ground under plants in the amount of 25–30 individuals from one local site. Nettle was also sampled from the local sites.

Specific activity of Sr-90 in the samples of soil, plants and terrestrial mollusks was measured on the scintillation beta spectrometer BETA 01C after a preliminary radiochemical isolation of the radionuclide. It was carried out following the routine procedure to determine Sr-90 in environment media through β-radiation of its daughter radionuclide Y-90. Measurements of the Sr-90 specific activity were repeated 3 times.

To calculate the absorbed dose rate on a mollusk with the Monte Carlo method, the MCNP5 program was applied.

## 3. Results and discussion

Measurements gave the following values of Sr-90 specific activities: 20±3 – 5203±785 Bq/kg in soil, 22.3±13.4 – 10596 ± 195 Bq/kg in plants, and 76±11 – 17640±2646 Bq/kg shells of terrestrial mollusk. These data were used as the input parameters for calculating the radiation dose rate.

### *3.1. Assessment of the radiation dose rate with the Levinger and Marinelli formulas*

This calculation is carried out taking into account the following sources of exposure of a mollusk.

The first source is a leaf of nettle. It is considered that a mollusk lives on a leaf during 155 days a year. The geometry of the irradiation source in this case is an infinitely thin plate. Radiation dose rate calculation is carried out according to the J. Levinger's formula [11].

The second source is soil, the geometry of which is regarded as an infinitely thin plate. For a warm period, when a mollusk lives on the soil surface, the radiation dose rate is also calculated by the J. Levinger's formula [11]. For the cold period (an average of 155 days), when a mollusk appears in an anabiosis and stays inside the soil, the dose rate is estimated from the L.D. Marinelli's formula [11].

The third source is a self-irradiation of a mollusk soft body from its shell. A shell is represented by the geometry of a sphere. The average parameters of mollusks inhabiting the study site are measured and taken into account. The average shell diameter is 1.4±0.1 cm, the average height – 0.8±0.1 cm, the radius – 0.72±0.01 cm, the square of a shell mouth – 6.16 ± 0.12 $cm^2$, and a shell volume of 1.44 ± 0.09 $cm^3$. To estimate the radiation dose rate on a mollusk soft tissue that is formed by external β-irradiation of Sr-90 and Y-90 containing in the shell, the formula is used for calculating the dose from the sphere radiation source to a point [11].

It is found that the radiation dose rate for a mollusk varies from 3.3±0.2 to 1634±245 mGy/year (table 1) along the local sites. At this, the radiation sources range in the following order: self-irradiation from shells (10±2 – 1189±178 mGy/year) > nettle (0.40±0.06 – 907±136 mGy/year) > soil (0.011±0.002 – 2.2±0.3 mGy/year). Important to emphasize that the largest contribution to the dose rate (over 95%) for a mollusk is brought by its self-irradiation from a shell.

### *3.2. Assessment of the radiation dose rate using the Erica Assessment Tool*

When assessing the radiation dose rate for a mollusk using the Erica Assessment Tool, only one source of irradiation could be taken into account – the soil. Calculated values of the radiation dose rates vary from 0.012±0.001 to 2.1±0.2 mGy/year (table 1).

**Table 1.** The radiation dose rate for a mollusk (mGy/year) calculated by different methods.

| Local site | J. Levinger and L.D. Marinelli formulas | Erica Assessment Tool | Monte Carlo method |
|---|---|---|---|
| 1 | 41.2±0.3 | 0.10±0.04 | 1.41±0.04 |
| 1a | 40.1±0.1 | 0.030±0.005 | 1.51±0.03 |
| 1b | 18.2±0.2 | 0.11±0.01 | 4.71±0.04 |
| 2 | 50.3±0.5 | 0.31±0.01 | 3.41±0.04 |
| 2a | 170±15 | 0.12±0.01 | 4.1±0.2 |
| 2b | 34.3±0.3 | 0.10±0.01 | 20.0±0.6 |
| 2c | 363±36 | 0.11±0.01 | 1.1±0.1 |
| 2d | 1287±124 | 0.010±0.001 | 61±9 |
| 3 | 19.3±0.9 | 0.12±0.01 | 4.1±0.1 |
| 3a | 912±43 | 0.031±0.001 | 49.2±0.5 |
| 3b | 842±36 | 0.027±0.001 | 47.1±1.4 |
| 4 | 32.1±2.6 | 0.10±0.01 | 1.5±0.1 |
| 4a | 512±27 | 0.052±0.001 | 24.3±0.7 |
| 4b | 611±24 | 0.061±0.001 | 26.0±0.7 |
| 4c | 499±19 | 0.053±0.001 | 25.0±0.6 |
| 6 | 105±24 | 0.31±0.01 | 2.7±0.1 |
| 6a | 188±31 | 2.1±0.2 | 14.0±0.4 |
| 6b | 648±85 | 0.21±0.02 | 32.0±0.7 |
| 6c | 1533±185 | 0.10±0.01 | 10.0±0.4 |
| 7b | 1634±245 | 0.9±0.1 | 76±9 |
| 10 | 378±81 | 0.23±0.04 | 11.1±0.5 |
| 10a | 160±34 | 0.9±0.1 | 18.0±1.3 |
| 10b | 76.2±14.3 | 0.011±0.001 | 3.1±0.1 |
| 10c | 45.1±19.3 | 0.12±0.02 | 1.23±0.04 |
| 10d | 30.0+11.2 | 0.12±0.01 | 1.10±0.05 |
| 10e | 156±24 | 0.11±0.01 | 9.0±0.3 |
| 10l | 327±84 | 0.32±0.02 | 12.1±0.3 |
| 11 | 667±94 | 0.9±0.2 | 10.0±0.6 |
| Control | 3.3±0.2 | 0.012±0.001 | 0.32±0.07 |

It should be noted that theses dose rates found with the Erica Assessment Tool are consistent with the results of the calculation with the D. Marinelli's formula for contribution of radioactive soil contamination (0.011±0.002 – 2.2±0.3 mGy/year). Despite the difference in the methods for calculating the radiation dose rate of a mollusk, the doses delivered from Sr-90 containing in soil are significantly correlated between two methods of estimation ($r = 0.96$).

### *3.3. Assessment of the radiation dose rate using the Monte Carlo method*

An application of the Monte Carlo method is recommended by the ICRP [6] for terrestrial ecosystems when both a source and a target could be located in soil or on its surface, unlike aquatic ecosystems where target organisms' density is close to the water density. The Commission note that numerical simulations of particle transport through natural environments implemented by the Monte Carlo method have certain advantages. They include a possibility to model complicated geometry of a target and a source, an adequate consideration of radiation physics, low inaccuracy in the assessments. This method is widely used in international projects in the field of the environment radiation protection to resolve tasks of dose assessments.

Since Sr-90 is the main dose-forming radionuclide at this territory, the solution of the dosimetric problem reduces to an estimation of the radiation transfer of the β-source of Sr-90 and the β-, γ-source of its daughter radionuclide Y-90.

For the radiation dose rate calculation, physical and geometric models of a mollusk are necessary to use as well as all possible scenarios of a mollusk exposure should be taken into account.

The physical model of a mollusk could be represented by the following parameters: the shell material of a mollusk, consisting of $CaCO_3$, has a density of 2.7 g/cm$^3$; soft shellfish is represented by a 4-component ICRU system with a density of 1 g/cm$^3$ [12, 13].

The geometric model has the following parameters:

- a mollusk is represented by the geometry of a sphere with a radius of 1 cm;
- a mollusk shell has the geometry of a spherical layer, the thickness of the layer is 25 * $10^{-4}$ cm;
- the weight of a mollusk is 7.2 g, which corresponds to an average weight of mollusks inhabiting the study territory;
- the weight of a mollusk shell is 2.7 g (an average weight of the shells of the mollusks inhabiting the study site).

The calculations performed by MCNP5 program allow for accounting for $10^9$ irradiation histories for each assumption. In calculations, nearly in all cases, a variance of less than 0.01% was obtained, in a few exceptional cases – less than 0.3%.

The following scenarios were considered at the assessment of the radiation dose rate for a mollusk.

Self-irradiation of an organism assumes the following assumptions:

- the source of radiation is deposited in a shell and soft tissue of a mollusk;
- the activities of Sr-90 in a shell and tissue are distributed in proportion to the Ca content in $CaCO_3$ (shell material) and in a mollusk tissue;
- the half-life of Sr-90 is not taken into account due to a short life cycle of a mollusk in comparison with the physical decay of the radionuclide.

The scenario when a mollusk lives on the soil surface assumes the following input parameters:

- the source of irradiation is enclosed in the volume of the soil, the geometry of the soil is represented by a cylinder;
- in the case of Sr-90, the cylinder has the following geometric parameters: radius – 10 cm, height – 6 cm; in the case of Y-90: radius – 14 cm, height – 8 cm. Taking into account different properties of Sr-90 and Y-90, different depth for the radionuclide deposition was used at the dose rate calculations;
- there is an air layer above the cylinder;
- the target (mollusk) is located in the center of the upper base of the cylinder.

The scenario of irradiation when a mollusk stays in anabiosis suggests the following:

- a mollusk is buried into the soil, the geometry of the soil volume is a cylinder with a radius of 19 cm and a height of 11 cm;
- a mollusk is located in the center of the cylinder along the radius and in the intermediate level (from 5 to 2 cm) along the height axis.

The scenario of irradiation when a mollusk is placed on nettle leaves during the vegetation period assumes the following input parameters and assumptions:

- the dose of irradiation is formed from the source of radiation distributed within the soil, the contribution to the dose from the source containing in the nettle is negligible;
- the target (mollusk) is at an altitude of 10 cm above the surface of the soil;
- for Sr-90 it is assumed that radiation is generated by the soil volume, represented by the geometry of the cylinder with a radius of 10 cm and a height of 10 cm;

- for Y-90, the radiation is assumed to be generated by the volume of the soil, represented by the geometry of two cylinders: with a radius and a height of 15 cm for β-radiation, and with a radius and a height of 20 cm for γ-radiation, respectively.

At further calculations of the radiation dose rate for a mollusk, the exposure time, animal weight and specific activity of radionuclides for all scenarios and at each local site were used.

As a result of all scenarios and assumptions, it is found that the radiation dose rate for a mollusk varies in the range from 0.32±0.07 to 76±9 mGy/year (table 1).

It is important to note that the radiation dose rate for a mollusk within nearly all the studied local sites is mostly formed from the self-irradiation of an animal that amounts to 0.3±0.02 – 48.6±0.4 mGy/year.

## 4. Conclusions

From the work done we can conclude that, among the methods considered, the most adequate estimates at the radiation dose rate calculations for a terrestrial mollusk could be received with the application of the Monte Carlo method. Such a conclusion is justified by the fact that the Monte Carlo method allows to take into account more possible scenarios of mollusk irradiation, maximum number of input parameters, including the weight and geometry of an organism, the dosimetric and physical features of a radiation source.

At the same time, the estimates obtained with the Erica Assessment Tool are the worst among the studied, because it is impossible to take into account an organism self-irradiation and contribution to the dose rate from radionuclides deposited in nettle leaves.

In turn, the assessment of the dose rate using the canonical formulas of J. Levinger and L.D. Marinelli allows considering all sources of an organism irradiation (soil, plant, self-irradiation). However, at this approach, such a very important input parameter as the weight of an exposed animal is not included in calculations. Since terrestrial mollusks are rather small biological organism with a low body weight (4–7 g), the weight omission can result in an overestimated value of the dose rate.

## References

[1] Brown J E, Alfonso B, Avila R, Beresford N A, Copplestone D, Pröhl G and Ulanovsky A 2008 *J. Environ. Radioact.* **99** 1371–83
[2] Larsson C M 2008 *J. Environ. Radioact.* **99** 1364–70
[3] Wood M D, Marshall W A, Beresford N A, Jones S R, Howard B J, Copplestone D and Leah R T 2008 *J. Environ. Radioact.* **99** 1484–95
[4] Vetikko V, Saxén R 2010 *J. Environ. Radioact.* **101** 82–7
[5] Antonova E V, Karimullina E M and Pozolotina V N 2013 *Russ. J. Ecology* **44** 18–27
[6] 2009 ICRP Publication 108. Environmental protection: the concept and use of reference animals and plants. *Ann. ICRP* p 242
[7] Kryshev A I, Sazykina TG, Strand P and Brown JE 2002 *Proc. 5th Int. Conf. on Environmental Radioactivity in the Arctic and Antarctic (Monaco)* 326–29
[8] Ulanovsky A, Pruhl G 2008 *Radiat. Environ. Biophysics* **47** 195–203
[9] Amiro B D 1992 *J. Environ. Radioact.* **35** 37–51
[10] Lavrentyeva G V 2014 *J. Environ. Radioact.* **135** 128–134
[11] Heine J and Brownell G 1958 *Radiation Dosimetry* (Moscow: Publishing house of foreign literature) p 637
[12] Briesmeister J F 2000 *MCNP A General Monte Carlo N-Particle Transport Code Version 4B* (M. Los Almos National Laboratory. Los Almos) p 741
[13] MCNP – A General Monte Carlo N-Particle Transport Code: Overview and Theory. 2003X-5 Monte Carlo Team. April. — Version 5

# About the development of a national system of response to the nuclear emergency for agriculture

**R A Mikailova and O A Shubina**

Russian Institute of Radiology and Agroecology, Obninsk, Russia

Email: mik_r_a@rirae.ru

**Abstract**. The document discusses the development of a system for responding to a nuclear emergency that could affect the agricultural sector. The analysis of some decision support systems showed that while the development of such a system there should be taken into account the advantages of the existing software tools allowing to derive recommendations for countermeasures and remediation of agricultural territories affected by the radiation accident. The consequences of major nuclear disasters have shown that the new system should consider the international experience on mitigation of the impact to agriculture, as well as consider potential nuclear emergencies, which can affect the territories located in the vicinity of nuclear facilities.

## 1. Introduction

Current tendencies in the development of nuclear power engineering suggest the decrease of the probability of emergency situations at nuclear power plants (NPP) and other enterprises of the nuclear fuel cycle (NFC) to the level of "inherent safety" [1]. Nevertheless, the disasters at the Chernobyl NPP and Fukushima Daiichi have shown the significance of the problem of nuclear accidents, and, as follows, there arises the necessity of the development of the plans for protective measures in case of different radioecological situations associated with the design-basis and beyond-design-basis accidents, which can result in the radioactive contamination of the environment [2].

Specialists consider three stages of the radiation accident. They are early (acute) stage, intermediate stage and late (remedial) stage.

Early stage is the period from the beginning of the accident until the moment, when the release of the radionuclides into the atmosphere stops and when the formation of the radioactive trail ends. The duration of this stage depends on the characteristics, accident scale and meteorological conditions; it can last from several hours to several days. At the early stage, the external radiation dose is formed due to the gamma- and beta-radiation from radionuclides contained in the cloud. Inhalation of radioactive products from the cloud causes the internal exposure of humans and animals. From the point of the radiation effect, radioactive noble gases make the significant contribution to the exposure of biological components of agricultural and natural ecosystems. Critical objects during this stage are the population, residing in the vicinity of the source of radioactive release, and agricultural animals, grazing on the territories affected by accident [3-5].

An intermediate stage of the accident begins from the moment, when the formation of the radioactive trail ends, and lasts until the implementation of all the necessary protective measures form the population and fulfilling of the required range of sanitary and hygienic, as well as medical and preventive measures. Depending on the characteristics and scale of the accident, the intermediate stage can last from several days to several months. During the intermediate phase the main reasons of harmful effects are external exposure to radioactive substances deposited from the cloud on the soil surface, buildings, etc., and internal exposure to the radionuclides because of their ingestion with drinking water and food products. The significance of inhalation is determined by the opportunity of breathing in the contaminated fine particles of soil, plant pollen, etc., resuspended by the wind. During this period the radioactive contamination of agricultural and natural ecosystems has already been formed. The concentration of radionuclides in plants decreases due to the radioactive decay, biomass buildup and crop purification under the meteorological factors [3-5].

The late (remedial) stage can take from several weeks to decades because accidental radioactive release can contain long-lived radionuclides. This stage ends when all the restrictions to the population activities on contaminated territories are cancelled and sanitary, and dosimetric control is performed regarding the radiation situation under the conditions of the "controlled exposure". The sources of external and internal exposure for this stage are the same as for the intermediate stage. The main way of radionuclide transport to agricultural migration chains is root uptake [3-6].

## 2. Consequences of the accidents at the Chernobyl and Fukushima Daiichi nuclear power plants for agriculture

Major radiation accidents have shown that agriculture is the main activity, impact to which determines total consequences of radiation accidents for humans.

We should mention that the Chernobyl accident was acknowledged as an "agrarian accident" due to the following facts:

- the region affected after the accident belonged to the zone of intensive agricultural production, where the agricultural sector is one of the leading sectors of the economy;
- the majority of contaminated territories belongs to the agricultural lands;
- the consumption of agricultural products is one of the main sources of additional exposure of population;
- the main population cohort are the people living in rural areas and having "rural" diet;
- doses of external and internal exposure of rural residents in the region of the accident are 1.3-4.0 times higher than those for the urban residents [7].

On the territories with high contamination levels, there were prohibited or significantly restricted traditional economic activities. According to the regulatory and legal acts in 1986-1991, there were excluded from agricultural use 2 640 km$^2$ of lands in Belarus, 171 km$^2$ in Russia and 1 583 km$^2$ in Ukraine. There were excluded from agricultural use 1 762.5 km$^2$ outwards the state radioecological conservation area in Belarus and 1 013 km$^2$ outwards the exclusion zone (30-km zone) in Ukraine. In Belarus, these lands are related to the zones of evacuation (exclusion) and top-priority resettlement, in Ukraine - to the zone of a mandatory evacuation, in Russia - to the exclusion zone [8-10].

Emergency on the Fukushima Daiichi NPP led to the contamination of vast areas, a significant part of which belong to agricultural and forest lands. Territories, which had to be remediated, were up to 500 km$^2$ (external dose higher 20 mSv/a) and 1 300 km$^2$ (external doses from 5 to 20 mSv/a). The investigations of the regions with the highest impact showed that contamination of paddy fields ranges from 67 to 41 400 Bq/kg, and contamination of other types of lands (arable, meadows, permanent crops) ranges from 16 to 56 600 Bq/kg. The most contaminated agricultural areas are located in Fukushima prefecture, where the content of radionuclides in 3.6% of all the samples (including 4% of samples from paddy fields and 2.9% of samples from other types of lands) exceed 5 000 Bq/kg [11].

## 3. Consideration of scenarios of potential accidents

Improvement of reactor facilities and nuclear fuel cycle, in general, is aimed at the heightening of economic effectiveness and radiation safety. The necessary condition of justification of new nuclear technologies is the assessment of the consequences of radiation impact to the environment resulted from the emergency situations at NFC enterprises. The reasons for such situations can have different nature: natural disasters, "human factor", terrorism and military actions, etc. Justification of the facilities of the nuclear industry, NPPs in particular, suggests not only the estimation of the radioactive effects on the environment during the normal operation but also in case of potential emergencies, including severe accidents [12].

The U.S. Nuclear Regulatory Commission initiated the State-of-the-Art Reactor Consequence Analyses (SOARCA) project to develop best estimates of the offsite radiological health consequences for potential severe reactor accidents for two pilot plants: the Peach Bottom Atomic Power Station in Pennsylvania and the Surry Power Station in Virginia. Peach Bottom is generally representative of U.S. operating reactors using the General Electric boiling-water reactor (BWR) with the gross capacity of 1412 $MW_e$ and a Mark I containment. Surry is generally representative of U.S. operating reactors using the Westinghouse pressurised-water reactor (PWR) with the gross capacity of 890 $MW_e$ and a large, dry (subatmospheric) containment [13, 14]. Accidental scenarios for Surry and Peach Bottom NPPs suggest the radioactive release of 60-70 different radionuclides with a total activity of up to $10^{19}$ Bq. Such situations can be rated $7^{th}$ level according to the International Nuclear Event Scale.

For example, within the conservative approach, the scenarios of the severe accidents caused by any seismic or other catastrophic external events consider the loss of power, failure of safety systems and core damage, and, as follows, radioactive release into the environment, estimated $3.36 \cdot 10^{18}$ Bq at a height of 8.4 meters in case of the accident at PWR-890 and $2.38 \cdot 10^{19}$ Bq at a height of 39.6 meters for the BWR-1412 (figure 1).

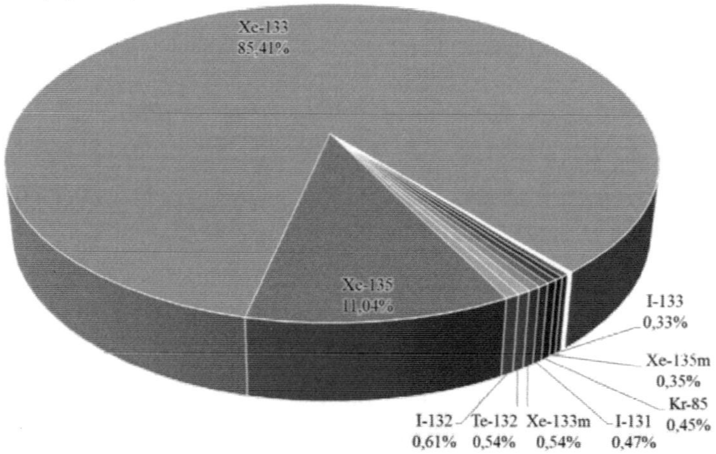


a)

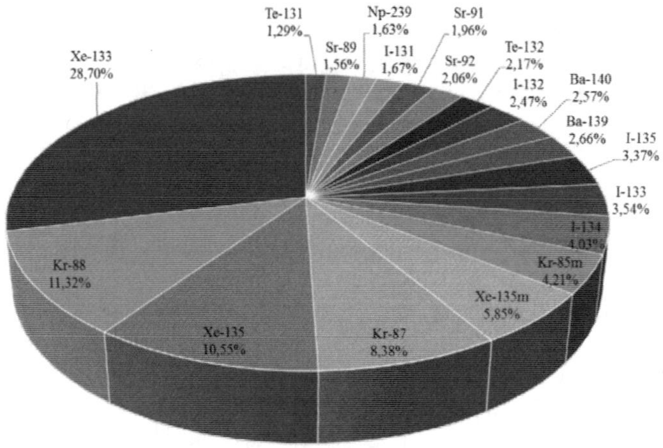


b)

**Figure 1.** Main radionuclides contained in the accidental radioactive release for NPPs with reactors PWR-890 (a) and BWR-1412 (b).

The analysis of scenarios [12] has shown that the accident at Surry NPP can result in the contamination (above 37 $kBq/m^2$) of ~90 $km^2$ of lands lying in the vicinity of the station, and in case of the accidental release from Peach Bottom NPP the contamination above 37 $kBq/m^2$ would cover ~150 $km^2$. These potential consequences might threaten the hypothetical agricultural activities performed on the territories located near such nuclear facilities. We should note that assessment of the radiation impact to the reference natural object (pine forest) showed that the affected area would be smaller: 4.5 $km^2$ for Surry and 100 $km^2$ for Peach Bottom. This confirms that fact that the territories with possible contamination of agricultural products would be larger than the areas with radiation damage of the natural ecosystems.

## 4. The idea of the national system responding to a nuclear emergency in the agricultural industry

In order to prevent and minimise the consequences of potential accidents at nuclear facilities for the agricultural sphere systems aimed at the response to the nuclear emergencies affecting food and agricultural production are necessary. Thus, the absence of the effective response system in agro-industry after the Chernobyl accident led to the unreasonably high content of $^{131}I$ in food products, consumed by the population and, as follows, to the increased entrance of $^{131}I$ into human organism and formation of high radiation doses for thyroid. This caused the significant number of thyroid cancers registered after the accident. In Japan, after the accident at Fukushima Daiichi NPP, they used the experience gained at the mitigation of the consequences after the Chernobyl accident. Actions taken during the emergency response in agriculture led to the reduction of radiation doses for the population during the first period after the accident to the relatively safe levels.

At the same time, not always reasonable actions during the emergency response in Japanese agriculture became the reason of formation of large volumes of radioactive wastes, i.e. to the generation of constant radioactive contamination sources. Moreover, this problem does not have any solution yet. Therefore, the timeliness and effectiveness of the measures to be taken during the emergency response in agro-industry directly influence the health of the population residing in the zone of the potential accident and possible damage to agricultural production due to radiation accidents. We should mention here, that nowadays the issues in the emergency response are of concern of international organisations, such as ICRP and IAEA. ICRP and IAEA issued new publications [15] and IAEA Safety Standards [16] containing requirements and recommendations, which are not considered now in terms of an existing emergency response system.

In the Russian Federation, there are response systems for the emergencies at facilities of the nuclear industry. For example, State Atomic Energy Corporation ROSATOM has the system of prevention and mitigation of emergencies [17]. Nevertheless, there is no national system of prompt intervention in the sphere of agricultural and food production in case of the radiation hazard, but at the same time there exist computer tools and decision-support systems, which could become the basis of such national response system. Current paper gives the example of some decision support systems (DSS) applied in the sphere of agriculture and food security:

PRANA. This system was developed in 1997 in Obninsk Institute for Nuclear Power Engineering. It is based on the principles and methods of radiological protection of the members of public and remediation of radioactively contaminated territories. It covers the basic countermeasures during the long-term period after the nuclear accident, their quantitative and qualitative characteristics including the rules and different restrictions of their application. This DSS uses GIS-technologies: vector maps of land use and associated databases, which are the system key components. During the assessment of the consequences of land contamination, recommendations on the application of protective measures and their efficiency, PRANA allows presenting the input/output information at different investigation levels (district, farms, fields, etc.) The particular attention is paid to agricultural measures: the assessment of their effectiveness is based on the use of economic and radiologic methods, including various kinds of cost-benefit analysis. Moreover, the program uses different methods for the analysis of uncertainties or inhomogeneous behaviour of population doses and contamination of agricultural products. Nowadays the system is developed for the contaminated territories of Bryansk region; maps and databases contain data on contamination levels of $^{137}Cs$ in soil and agricultural products [18].

ReSCA (Remediation Strategies after the Chernobyl Accident) was being developed within the IAEA TC projects in 2003-2008. It was aimed at the decision support in the sphere of population safety on contaminated territories in the Republic of Belarus, the Russian Federation and Ukraine to facilitate a selection of optimised remediation strategies in rural settlements. The DSS is based on data on contamination of agricultural products, settlements and pasture areas. Besides settlement-specific data, the system uses a set of generic country-dependent parameters, which are based on the expertise level knowledge gained during remediation activities in the affected countries. The principal output of ReSCA is an optimised strategy of remediation of contaminated areas at settlements, but there is a lack of recommendations on performing of activities on agricultural lands. As the input data, the program uses the values of concentrations of $^{137}Cs$ in soil and doesn't suppose any geoinformation units [19].

Also, we should mention here that decision support systems PRANA and ReSCA were developed before the Fukushima accident and, as follows, they do not consider the experience on mitigation of consequences and various parameters gained during the studies, which were conducted by the international community after 2011.

The RODOS system is the result of a close collaboration between almost 40 institutes from about 20 countries within the European Union, Eastern Europe and the former Soviet Union. The development of the system began at the end of 1980-s. Since 2010 RODOS system was followed by its newest JAVA based version named JRodos, which has been improved until the present time. JRodos is a decision support system for off-site emergency management and remediation issues after nuclear accidents, which has the primary aim to provide consistent and comprehensive information at local, regional and national levels, during all stages of a real event and while preparing for possible future events. The system has a module for deriving the recommendations on the application of agricultural countermeasures during the long-term period after the accident. The model estimates the effectiveness of the strategy according to the total contamination of food products, radiation doses, required sources, wastes and costs taking into consideration the time span during which pre-defined activity levels in food would be exceeded. The assessment is performed within the application of countermeasures in production and consumption of food products. The program considers the following countermeasures: food disposal and stopping of food production, food processing and storage, removal of animals from contaminated feed, use of sorbents in animal feed, change of crop species or variety, amelioration and changes in the types of land use [20, 21].

The development of Decision Support System for Nuclear Emergencies Affecting Food and Agriculture (DSS4NAFA) started in 2013 within the framework of the FAO/IAEA Coordinated Research Project D15015 "Response to Nuclear Emergencies Affecting Food and Agriculture". This decision support system has been developed and is being improved on the basis of contemporary information technologies for the management of data collection and visualisation and decision-making to ensure food safety and security using well-timed and complex planning on restrictions in food and agriculture production during a nuclear emergency. Beta version for extensive independent testing is expected to be available in July 2018. The DSS allows performing tasks on the response to nuclear emergencies considering the gained experience and situation studies. DSS4NAFA can be used during both emergency responses as well as for routine monitoring of nuclear and radiological incidents that could affect food and agriculture. Data here is managed in real-time and up-to-date information on the spatial and temporal distributions of radionuclides in the landscape for decision-makers and end-users [22, 23].

## 5. Conclusions

Development of the national nuclear emergency response system for agriculture should consider the international experience on mitigation of the consequences of nuclear accidents and use the advantages of the decision support systems mentioned above as its basis, taking into the consideration the regulatory documents and standards of the Russian Federation. The analysis of these DSS has shown that to organise the efficient agricultural production regarding an emergency, the recommendations on countermeasure implementation should consider particular soil characteristics, peculiarities of separate farms and positive and negative experience of managing the effects resulted from major nuclear disasters. Moreover, such system should take into account the potential accident scenarios for the nuclear power plants located in the Russian Federation because the damage and consequences depend not only on the external environmental conditions but reactor types, main radionuclides and parameters of the radioactive release.

In summary, the national system of response to nuclear emergency in the sphere of agricultural production should include the advantages of contemporary decision support systems, which allow to derive recommendations on the performing of agricultural activities during the nuclear emergency for different investigation levels (region, farm, field, etc.) and monitoring of contaminated farmlands with the consideration of economic, agricultural and social aspects of the considered territories.

**References**

[1] Adamov E O and Solovyov D S 2017 *The Energy Policy* **3** 21–30
[2] Alexakhin R M 2013 *Atomic energy* **114** 243–9
[3] UN 2011 *Sources and Effects of Ionizing Radiation: UNSCEAR 2008 Report* vol. 2 (Vienna: UN Publishing and Library Section) p 179
[4] Alexakhin R M and Sirotkin A N 2001 Chernobyl disaster and agrarian science *Chernobyl. Duty and Courage: Scientific and journalistic monograph* (Moscow: the 4th branch of Voenizdat) pp 474–506
[5] Vladimirov V A and Izmalkov V I 2000 *Disasters and Ecology* (Moscow: Kontakt-Kultura) p 382
[6] Sanzharova N I *et al.* 2016 *Radiatsionnaya Biol. Radioekol* **56** 322–35
[7] Alexkhin R M, Sanzharova N I, Fesenko S V, Spirin E V, Spiridonov S I and Panov A V 2006 *Chernobyl, agriculture, environment: Proc to the 20$^{th}$ anniversary of the accident at the Chernobyl nuclear power plant in 1986* (Obninsk: RIARAE) p 35
[8] 2009 *Atlas of current and forecast effects of the Chernobyl accident on affected parts of the Russian Federation and Belarus (ACFE Russia-Belarus)* ed Yu A Israel and I M Bogdevich (Moscow: Foundation «Infosfera» – NIA – Priroda) p 140
[9] Israel Yu L 2006 *Radioactive contamination of natural environments resulted from the Chernobyl nuclear power plant accident* (Moscow: Koltekhprint) p 28
[10] Shubina O A, Titov I E and Krechetnikov V V 2016 Remediation of the lands temporarily excluded from land use after the Chernobyl accident *Chernobyl - 30 years later. Radiation and hygienic aspects of negotiating the consequences of the Chernobyl accident: Proc. of the int. research and practice conf.* (St. Petersburg: RIRH) pp 200–2
[11] Bachev H and Ito F 2017 *Ann Agric Crop Sci* **2** 1030
[12] Mikailova R A and Spiridonov S I 2018 *Atomic Energy* **123** 165–70.
[13] U.S. NRC 2013 *State-of-the-Art Reactor Consequence Analyses Project Volume 1: Peach Bottom Integrated Analysis (NUREG/CR-7110)* vol 1 rev 1 (Washington: Office of Nuclear Regulatory Research) p 329
[14] U.S. NRC 2013 *State-of-the-Art Reactor Consequence Analyses Project Volume 1: Peach Bottom Integrated Analysis (NUREG/CR-7110)* vol 2 rev 1 (Washington: Office of Nuclear Regulatory Research) p 565
[15] ICRP Publication 103 2007 The 2007 Recommendations of the International Commission on Radiological Protection. *Ann. ICRP* **37** p 332
[16] IAEA 2017 *Preparedness and response for a nuclear or radiological emergency: general safety requirements No. GSR Part 7* (Vienna: IAEA) p 102
[17] SC “Rosatom” 2017 *The results of the activities of the State Atomic Energy Corporation "Rosatom" for 2016: Public annual report* (Moscow: SC “Rosatom”) p 183

IOP Conf. Series: Materials Science and Engineering **487** (2019) 012013 doi:10.1088/1757-899X/487/1/012013

[18] Yatsalo B, Mirzeabassov O, Okhrimenko I, Pichugina I and Kulagin B 1997 *Radiat Prot Dosimetry* **73** 291–4
[19] Ulanovsky A, Jacob P, Fesenko S, Bogdevitch I, Kashparov V and Sanzharova N. 2011 *Radiat Environ Biophys* **50** 67–83
[20] Raskob W, Trybushnyi D, Ievdin I, Zheleznyak M 2011 *Radioprotection* **46** 731–6
[21] KIT 2017 *JRodos: An off-site emergency management system for nuclear accidents* (Karlsruhe: KIT) p 22 (https://resy5.iket.kit.edu/JRODOS/documents/JRodos_Report_forHomepage.pdf)
[22] Shubina O A and Mikailova R A 2016 *Medical radiological consequences of the Chernobyl accident: the forecast and the actual data after 30 years* (Obninsk: Tsyb MRRC) p 191
[23] FAO/IAEA 2018 *Response to Nuclear Emergency affecting Food and Agriculture - D1.50.15* (http://www-naweb.iaea.org/nafa/swmn/crp/swmcn-nuclear-emergency-food.html)

IOP Conf. Series: Materials Science and Engineering **487** (2019) 012014 doi:10.1088/1757-899X/487/1/012014

# Technical approaches to calculating different components of the annual effective dose due to release of actinides into air during long-term operation of a BN-type reactor

**O.O. Peregudova, A.G. Tsikunov**

State Scientific Centre of the Russian Federation – Institute for Physics and Power Engineering named after A.I. Leypunsky, Obninsk, Russia

e-mail: operegudova@ippe.ru.

**Abstract**. The aspect of ensuring NPP safety has always attracted scrupulous attention. Experience in operating power plants with different reactor types shows that radiation situation in the NPP area, environmental contamination depend very much on the integrity of the fuel cladding – one of the most important protective barriers.

In practice, some fuel pins in the core can have certain cladding defects occurring during the normal plant operation, with a BN-type reactor used. Np, Pu, Am, Cm transuranic elements may take special part in contaminating the primary circuit with actinides. Presence of actinides in the process media of the reactor facility may have radiation effect not only on the NPP staff but also on the population living outside the buffer area.

There is little information on the problem in scientific publications. Computer codes are mainly oriented to the calculation of the effective doses the population take as a result of release of radionuclides into the air during the first year of NPP operation. Increasing the duration of NPP operation results in significantly complicated calculations. This paper presents a simplified method of calculating the main components of the annual effective dose the population receive due to release of actinides into the air over «n» partial refuelings after reactor startup.

## 1. Introduction

The aspect of ensuring NPP safety has always attracted a scrupulous attention. As early as at the stage of designing first power producing fast reactors, much attention was paid to the safety aspect, both during their normal operation and in emergency situations. Search directions for corresponding design solutions were dictated by the requirement to eliminate intolerable impact on the environment and population by means of reactor inherent safety, effective systems of accident management mitigating the aftermath.

Experience in operating power plants with different reactor types shows that radiation situation in the NPP area, environmental contamination depend very much on the integrity of the fuel cladding – one of the most important protective barriers.

In practice, some fuel pins in the core can have certain cladding defects occurring during the normal plant operation, with a BN-type reactor used. These are microcracks that cause fission gas release or macrocracks that can lead to direct interaction of sodium with the fuel, with the result that a certain amount of nonvolatile fission products or even fuel may appear in the sodium. In the latter case, Np, Pu, Am, Cm transuranics may play special part in contaminating the primary circuit with actinides.

Presence of actinides in the process media of the reactor facility (in the primary-circuit sodium, on the inner surfaces of the primary-circuit equipment that cannot be removed, on the surfaces of the SFA discharged from the reactor facility, etc.) may have radiation effect not only on the NPP staff but also on the population living outside the control area.

There is little information on the problem in scientific publications. The known computer codes RELEASE-3.1 (Certification ID 395 of July 14, 2016), EXPRESS [1]) are mainly oriented to the calculation of the effective doses the population take as a result of release of radionuclides into the air during the first year of NPP operation. Increasing the duration of NPP operation results in significantly complicated calculations. This is most typical of long-lived actinides and is the focus of research for the authors [2, 3].

The paper presents a simplified method of calculating the main components of the annual effective dose the population receive due to release of actinides (for the specific $i$ – th radionuclide) into the air over «n» partial refuelings after reactor startup:

- The annual effective dose of external exposure due to actinide cloud – $D_{cloud}(n)$.
- The annual effective dose of external exposure due to surface contamination of soil with actinides – $D_{surface}(n)$.
- The annual effective dose due to actinide inhalation – $D_{inhalation}(n)$.
- The annual effective dose due to food chains (via air) - $D_{food}^{air}(n)$.
- The annual effective dose due to food chains (via roots) - $D_{food}^{root}(n)$.

Total effective dose is a sum:

$$D_{total}^{i}(n) = D_{cloud}^{i}(n) + D_{surface}^{i}(n) + D_{inhalation}^{i}(n) + D_{i,food}^{air}(n) + D_{i,food}^{root}(n)$$

The evaluated components of the effective dose are the result of added contributions of all the actinide isotopes involved in the calculations. Presented below are algorithms for calculating specific components of the effective dose. The reactor facility is supposed to operate in a steady state mode.

## 2. Methods and results

### 2.1. Annual effective dose of external exposure due to actinide cloud

The annual effective dose of external exposure due to actinide cloud is estimated as

$$D_{cloud}^{i}(\mathrm{n}) = D_i^I \cdot K_i^{circuit}(\mathrm{n})$$

where

$D_i^I$ is the dose of external exposure due to actinide cloud. It is the average individual dose the population receive (from the cloud) for a particular $i$ – th radionuclide at the early stage of BN reactor operation (dose over the first year of reactor operation). The RELEASE-3.1 code is used for this purpose at SSC RF-IPPE;

$K_i^{circuit}(n)$ is the coefficient to include the change in the content of i-th radionuclide in the primary circuit. For each nuclide, it is necessary to find the dependence of the $K_i^{circuit}(n)$ value on the time of BN reactor operation (it marks the increased effective dose the population receive over time).

$$K_i^{circuit}(n) = \frac{[1-(1-\omega)^n \cdot e^{-(\Delta\tau+\tau)\cdot\lambda_i\cdot n}]}{[1-(1-\omega)\cdot e^{-(\Delta\tau+\tau)\cdot\lambda_i}]}$$

where

$\omega$=0,05 is the annual fraction of fuel that escapes from the primary circuit when a spent FA is reloaded (from fuel present in sodium and on the inner surfaces of primary-circuit equipment);

$\tau$ – is the duration of the partial refueling, eff. days;

$\Delta\tau$ – is the duration of the refuelling shutdown;

($\Delta\tau$+$\tau$)=(35 days+330 eff. days);

$\lambda_i$ – is the decay constant of the $i$ – th radionuclide;

$N$ – is the number of fulfilled partial reloadings.

*2.2. Annual effective dose of external exposure due to surface contamination of soil with actinides*

The annual effective dose of external exposure due to surface contamination of soil with actinides is estimated from

$$D_{surface}(n) = (1-0{,}004)\cdot D_i^{II}\cdot K_i^{circuit}(n)\cdot K_i^S(n);$$

Based on experience and conducted research, the following correction is introduced into the formula: 0,004 = 0,4% is the average amount of drain per year (share of the i-th radionuclide that penetrated into soil surface).

The authors will calculate the $D_i^{II}$ value by using the RELEASE-3.1 code. The value is the average individual dose the population receive (from the contaminated surface) at the early stage of BN reactor operation (dose over the first year). The $K_i^{circuit}(n)$coefficient to include the change in the amount of actinides in the primary circuit («n» dependence) has already been used in Section 1.

$$K_i^{circuit}(n) = \frac{[1-(1-\omega)^n\cdot e^{-(\Delta\tau+\tau)\cdot\lambda_i\cdot n}]}{[1-(1-\omega)\cdot e^{-(\Delta\tau+\tau)\cdot\lambda_i}]}$$

$K_i^S(n)$ is the coefficient which allows a more accurate estimation of the dynamics of change in the content of $i$ – th radionuclide in the contaminated soil layer 5 cm thick during long-term BN reactor operation (5 cm is the surface root soil layer [4]).

$$K_i^S(n) = \frac{(1-e^{-(\lambda^i_{discharge}+\lambda^i_{migration}+\lambda^i_{decay})\cdot(\tau+\Delta\tau)\cdot n})}{(1-e^{-(\lambda^i_{discharge}+\lambda^i_{migration}+\lambda^i_{decay})\cdot(\tau+\Delta\tau)})}$$

where

$\lambda^i_{discharge}$ is the constant of washing the $i$ – th radionuclide out of the contaminated soil layer 5 cm thick;

$\lambda^i_{migration}$ is the constant of $i$ – th radionuclide release from the contaminated soil layer 5 cm thick due to migratory processes.

The indicated constants ($\lambda^i_{discharge}$, $\lambda^i_{migration}$) are presented in paper [4] and amount to $\lambda^i_{discharge}$ = 0,006 /year$^{-1}$; $\lambda^i_{migration}$ = 0,122 /year$^{-1}$.

*2.3. Annual effective dose due to actinide inhalation*

The annual effective dose due to actinide inhalation is calculated from

$$D^i_{inhalation}(n) = D_i^{III}\cdot K_i^{circuit}(n)\cdot K_i^{DECAY,I}(n),$$

where

$D_i^{III}$ is the average individual dose the population receive due to actinide inhalation at the early stage of BN reactor operation for a year, calculated by using the RELEASE-3.1 code;

The $K_i^{circuit}(n)$ coefficient has already been used in Sections 1 and 2. Unchanged, the coefficient will be used and considered in Section 3.

Another coefficient - $K_i^{DECAY,I}$ - is to be introduced to include biological decay of the inhaled radionuclides.

$$K_i^{DECAY,I} = \frac{(1-r^n)}{(1-r)},$$

where

$$r = e^{-(\lambda_{decay}^i + \lambda_{biol.dec.inh.}^i)\cdot(\tau+\Delta\tau)};$$

$$r^n = e^{-(\lambda_{decay}^i + \lambda_{biol.dec.inh.}^i)\cdot(\tau+\Delta\tau)\cdot n};$$

$\lambda_{biol.dec.inh.}^i$ is the constant of biological decay of the i-th inhaled radionuclide due to biological processes.

The indicated constant is presented in paper [5] and amounts to 0,693/25 years=0,02772 year$^{-1}$.

*2.4. Annual effective dose due to food chains (via air)*

The annual effective dose due to food chains (via air) is calculated as following

$$D_{i,food}^{air}(n) = D_i^{IV} \cdot K_i^{circuit}(n) \cdot K_i^{DECAY,I}(n),$$

where

$D_i^{IV}$ is the annual effective dose the population take due to food chains (via air) after the first year of BN reactor operation, calculated by using the RELEASE-3.1 code;

The $K_i^{circuit}(n)$ coefficient has been used in Sections 1, 2 and 3 (it is proposed to be used unchanged in Section 4);

The $K_i^{DECAY,I}(n)$ coefficient has been used in Section 3 (it is proposed to be used unchanged in Section 4).

*2.5. The annual effective dose due to food chains (via roots)*

The annual effective dose due to food chains (via roots) is calculated as

$$D_{i,food}^{root}(n) = D_i^{V} \cdot K_i^{circuit}(n) \cdot K_i^{DECAY,II}(n),$$

where

$D_i^{V}$ is the annual effective dose the population take due to food chains (via roots) after the first year of BN reactor operation, calculated by using the RELEASE-3.1 code;

$K_i^{circuit}(n)$ has been used in Sections 1, 2, 3 and 4 (it is proposed to be used unchanged in Section 5);

In case of the «root» variant, a new $K_i^{DECAY,II}(n)$ coefficient is introduced.

$$K_i^{DECAY,II}(n) = \frac{(1-r^n)}{(1-r)},$$

where

$$r = e^{-(\lambda^{i}_{discharge}+\lambda^{i}_{migration}+\lambda^{i}_{decay}+\lambda^{i}_{\substack{biol.dec.\\(via\\roots)}})\cdot(\tau+\Delta\tau)},$$

$$r^{n} = e^{-(\lambda^{i}_{discharge}+\lambda^{i}_{migration}+\lambda^{i}_{decay}+\lambda^{i}_{\substack{biol.dec.\\(via\\roots)}})\cdot(\tau+\Delta\tau)\cdot n}$$

The indicated constant of ( $\lambda^{i}_{\substack{biol.dec.\\(via\\roots)}}$ ) is presented in papers [4, 5].

## 3. Conclusions

This paper presents a simplified method of calculating the main components of the annual effective dose the population receive due to release of actinides (for the specific $i$ – th radionuclide) into the air over «n» partial refueling after the startup of a reactor with defective fuel claddings in the core.

Further calculations are planned to find the radiation effect of actinides on the population under long-term operation of an NPP with a BN-type reactor. These data are crucial for designing BN reactor facilities where fuel with a high content of transuranic Pu, Np, Am, Cm elements will be used

## References

[1] The methodology for calculating environmental pollution and assessing the impact on the population to justify protection measures in the event of an accident at radiation hazardous facilities with the Express software, MBR 6.2.15-09, 2009

[2] Peregudova O O, Tsikunov A G, Seliverstov I N and Mukhamedeev R I 2014 *Radiation impact of actinides on the population in the normal operation of nuclear power plants with a BN-type reactor. Questions of atomic science and technology*. Series: Nuclear-reactor constants. **4** 14–23

[3] Tsikunov A G, Seliverstov I N, Mukhamadeev R I and Peregudova O O Assessment of the contribution of actinides to the radiation effect on the population in the normal operation of a nuclear power plant with a high-power energy reactor on fast neutrons with a sodium coolant 2014 *Radiation and Risk (Bulletin of the National Radiation and Epidemiological Register)* **23** 81–91

[4] Alexakhin R M 1982 *Nuclear Energy and the Biosphere* (Moscow: Energoizdat) p 216

[5] Buldakov L A 1990 *Radioactive Substances and Man* (Moscow: Energoatomizdat) p 160

# Transfer of $^{137}$Cs into onion (*Allium cepa L.*) under conditions of hydroponic model experiment

**M M Godyaeva[1], T A Paramonova[1] and N V Kuzmenkova[2]**

[1]Department of Radioecology and Ecotoxicology, Faculty of Soil Science, Moscow State University, Moscow, Russia
[2]Department of Radiochemistry, Faculty of Chemistry, Moscow State University, Moscow, Russia

Email: airrune@yandex.ru

**Abstract**. Short-term hydroponic experiment was conducted to evaluate intensity of the $^{137}$Cs transfer from a solution with extremely high activity of the radionuclide into onion, considering different plant's compartments (green leaves, bulbs, roots). It has been revealed that the root uptake of $^{137}$Cs was significantly discriminated, and not exceeded 0.01% of the total amount of radionuclide contained in the solution. At the same time, biometric indices of the plants – shoot height, roots elongation, phytomass inventory and structure – under conditions of $^{137}$Cs stress exposure were not significantly changed in comparison with the control. The main accumulation of $^{137}$Cs, which incorporated into onion, occurred in roots, especially in root tips, which are most actively involved in the processes of mass exchange between the plant and the external environment. The translocation of $^{137}$Cs into the shoot part, including the onion bulb and green leaves, was significantly suppressed, which allows considering the root system of onion as a biological barrier providing the rhizofiltration of the toxicant.

## 1. Introduction

Global contamination of terrestrial ecosystems of the Northern Hemisphere by $^{137}$Cs started from the 50s years of the 20th century as a result of nuclear weapons tests in the atmosphere, and after limiting and prohibiting such tests by the international pact of 1963 and 1996, introduced into the biosphere anthropogenic radioactive elements was associated with major accidents at nuclear power facilities. For example, the release of technogenic radionuclides into the atmosphere at Kyshtym accident (1957, USSR) was estimated as $7.4 \cdot 10^{16}$ Bq, Windscale/Sellafield accident (1957, UK) – $7.7 \cdot 10^{14}$ Bq, Chernobyl accident (1986, USSR) – $5.3 \cdot 10^{18}$ Bq, Goiania accident (1987, Brazil) – $5.1 \cdot 10^{13}$ Bq [1]. Emissions into the atmosphere as a result of the accident at NPP Fukushima Daiichi are estimated to be approximately one tenth of the emissions from the Chernobyl accident [2]. The ecological consequence of the distribution of long-lived radionuclides in the environment is their inevitable introduction into the biological cycle of elements, total, with their root consumption by plants from contaminated soils, and in the future - when spreading through food chains, the end user of which is possibly a human. The physiological role in living organisms of the natural analog of $^{137}$Cs, stable $^{133}$Cs, is currently unclear. However, it is presumably capable of causing metal toxicity stress, since it belongs to a group of heavy, and according to some ideas, even superheavy metals [3]. In turn to the incorporation of $^{137}$Cs into living organisms can complement the negative effects of chemical nature by the physical impact of ionizing radiation on them. Passing to a stable state, $^{137}$Cs undergoes beta decay, which produces a barium isotope $^{137}$Ba ($T_{1/2}$ 30.17), with 94.4% of the decay occurring with the intermediate formation of the $^{137}$Ba nuclear isomer (T1 / 2 2.55 min) , which in turn goes to the ground state with the emission of a γ-quantum with an energy of 661.7 keV. The total energy released during the decay of one $^{137}$Cs nucleus is 1175.63 ± 0.17 keV. Value of $^{137}$Cs as the main dose-forming radionuclide of bomb-derived and accidental fallout is amplified by the duration of its half-life and the ability to be firmly and irreversibly fixed by clay minerals of soils in the uppermost layer, while at the same time the main part of plant roots is concentrated [4, 5]. Thus, the greater $^{137}$Cs entering terrestrial ecosystems is still in the space of intensive mass exchange between soil and plants, which facilitates the reproduction of the biogeochemical cycle of the radionuclide [4, 6, 7]. Hence, radioactive contamination of terrestrial ecosystems by $^{137}$Cs is a matter of the recent environmental concern. Numerous studies on the intensity of $^{137}$Cs root consumption by different plant cultures have shown that the specific activity of $^{137}$Cs in plants is, as a rule, much less than in soils, i.e. bioavailability of the radionuclide in the field conditions is rather low [8]. However, in the mitigation of the $^{137}$Cs root consumption process, it is difficult to differentiate the role of soil characteristics (namely, immobilization of the radionuclide due to irreversible sorption, as well as the configurational inaccessibility of the element localized inside the cloddy waterproof aggregates, etc.) and the actual biological properties of plants. The last once can be detected under conditions of hydroponic experiment, when the radionuclide is in ionic form and is completely accessible to plants. Besides, in comparison with field conditions, hydroponic experiments are easy to control and require less time to get general conclusions. Thus, they are used to investigate the bioaccumulation processes of potential phytotoxicants [9, 10]. Based on this, the purpose of this study was to analyze the intensity of the $^{137}$Cs transition to onion plants (*Allium cepa* L.) and to identify the features of translocation efficiency from root to shoot in conditions of soilless hydroponic culture cultivation. In turn, the understanding of sensitivity or resistance of the crop to potentially phytotoxic radionuclide, as well as the revealing of mechanisms of biological adaptation to environmental stress are the vital factors in choosing suitable plants to grow at radioactively contaminated areas.

## 2. Materials and methods

### *2.1. Plant material and hydroponic experiment design*

Short-term indoor hydroponic experiment was carried out with onion bulbs of commercially obtained Stuttgart Riesen cultivar. Onion has been chosen due to big supplies of reserve nutrients that create the possibility of growing plants without physiological nutrient solution to avoid competition effects of monovalent beneficent ions ($K^+$, $NH_4^+$). Besides, onion is known as a plant with sensitivity to toxicants, and by the opinion of A.A. Oudalova et al. [11] special-purpose *Allium*-test could be recommended as an effective tool for mutagenic, mitosis-modifying and toxic testing in case of combined contamination of radioactive and chemical origin.

7 onion bulbs were placed in a plastic box with $^{137}$CsCl solution with extremely high activity of 2.8 MBq/l and were grown for 14 days at day and night temperatures of 23 °C and 18 °C, respectively, with 45–60% relative humidity, and 250 μmol/m$^2$/s light intensity. At the same time, a control experiment was conducted on the cultivation of onion on deionized water.

The vital state of plants was assessed; the growth of shoots and the length of leaves of onion were recorded daily.

### *2.2. Sampling and laboratory analysis*

After 14 days of vegetation the onion plants were removed from the box, and were carefully rinsed with deionized water. The root, bulb and leaf parts (the last – at the base of a stem) were separated using a scalpel. Fresh weight of the biomass compartments was evaluated.

Then 2 copies of the onion plants grown in the $^{137}$Cs-containing solution were herbicated (have been pressed and dried) for further autoradiography studies. Remaining plants were dried at a temperature of 80 °C in an oven until completely dried, were ground using a mortar and pestle for homogenization of plant material, and prepared for the further γ-spectrometric analysis.

The prepared plant samples were studied using digital autoradiography on a Perkin Elmer Cyclone device (USA). The samples were isolated from the external γ-background using a shelter made of lead plates. Since the $^{137}$Cs activity was different, the exposure times were individually selected; on average, it was 24 hour.

Spectrometric measurements were performed using a GR 3818 semiconductor γ-spectrometer with a Canberra high-purity (HPGe) detector (United States). Before the start of the measurements, three calibration of the γ spectrometer with a plant standard of IAEA No. 372 with a mass of 0.5, 1 and 1.5 g were carried out. The registration efficiency coefficients were 0.06, 0.05 and 0.05, respectively. The specific activity of $^{137}$Cs in prepared onion samples was measured in the standard geometry of the Petri dish. Energy spectra measurements were accounted for 1200 sec, and $^{137}$Cs activities were calculated from the net full energy peak 661.7 keV using the "Progress 5.1" spectrometric analysis software package. The analytical error did not exceed 2%.

### *2.3. Calculations and statistical analysis*

The tolerance index (TI) to estimate generally the process of plant growth and biomass yield was calculated by the equation:

$$TI = \frac{M_{treatment}}{M_{control}},$$

where $M_{treatment}$ is total dry mass of onion or the mass of the plant compartments under $^{137}$Cs-containing hydroponic solution, and $M_{control}$ is dry mass of the plants and their parts in control.

To assess intensity of $^{137}$Cs transfer from hydroponic solution into onion plants the values of transfer factor (TF) were calculated:

$$TF = \frac{A^{Cs-137}{}_{plant}}{A^{Cs-137}{}_{solution}},$$

where $A^{Cs\text{-}137}{}_{plant}$ is specific activity of $^{137}$Cs in dry mass of plant, $A^{Cs\text{-}137}{}_{solution}$ is specific activity of $^{137}$Cs in hydroponic solution.

The specific bioaccumulation coefficient (BC) for the onion compartments was calculated as:

$$BC = \frac{A^{Cs-137}{}_{plant\ compartmen}}{A^{Cs-137}{}_{solution}},$$

where $A^{Cs\text{-}137}{}_{plant\ compartment}$ (roots, bulbs, leaves) is specific activity of $^{137}$Cs in dry mass of the once.

To quantify the translocation of $^{137}$Cs from roots to aerial parts the translocation coefficient (TC) was used as:

$$TC = \frac{A^{Cs-137}{}_{bulbs\ or\ leaves}}{A^{Cs-137}{}_{roots}},$$

And the last, but not the least, potential phytoremediation factor (PPF) was defined as:

$$PPF = \frac{A^{Cs-137}{}_{plant} \cdot M_{plant}}{A^{Cs-137}{}_{solution} \cdot V_{solution}} \cdot 100\,(\%),$$

where $M_{plant}$ is total dry mass of plant, $V_{solution}$ is total volume of hydroponic solution.

To assess mean values and variability of the obtained data Microsoft Excel 2010® (Microsoft Cooperation, USA) was performed.

## 3. Results and Discussion

### 3.1. Plant growth and biomass yield

After 14 days of the onion plants exposure to $^{137}Cs$ influence, there were no obvious impact on the aboveground parts and roots. The most significant toxic symptoms, such as chlorosis and senescence of leaves, wilted leaf tips, and changes in color (blackening) of roots were not noted in comparison with the control.

Changes in growth variables throughout plant development under $^{137}Cs$-stress conditions were not significant comparing to control for the first week, but slightly raised during the second week of the experiment (figure 1). As in the experiment and in the control on the 12$^{th}$-13$^{th}$ day of hydroponic cultivation of onion, the growth of leaves in height slowed down and passed to the stage of the plateau. By the end of the experiment, plants grown in the $^{137}Cs$-containing solution had a slightly less developed habit than in the control variant, but these differences were not statistically significant. One of the 7 bulbs used in the experiment began to germinate only after the 10th day of the experiment, and to its completion the height of the shoot was only 2.9 cm, which was a > 5 time less than in the control. In the control, the activation of plants from the dry bulb condition to the vegetative growth phase was 100%.

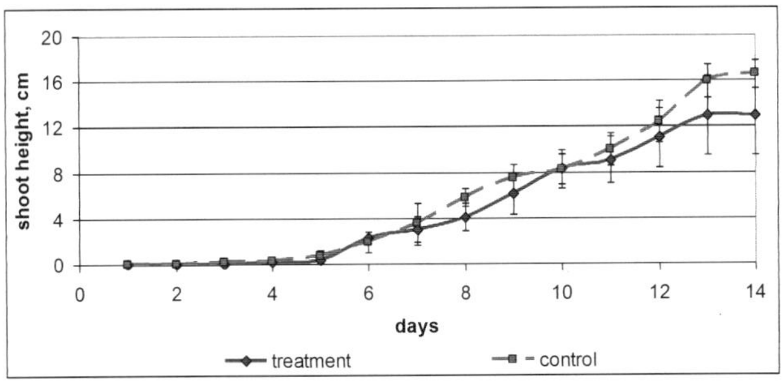


**Figure 1.** Dynamics in onion shoot growth during hydroponic experiment under $^{137}Cs$ treatment and in control conditions. Error bars represent $\pm t_{0.95} \cdot m$ at n = 7.

During the experiment, the raw weight of plants increased by ≈ 40%. At the same time, the height of shoots and root length, total biomass yield and the structure of its stocks by plant compartments when growing onion in the $^{137}Cs$-containing solution were close to the control parameters (table 1), i.e. the significant phytotoxic effect of the radionuclide on biometric indices of onion in experimental and control conditions has not been established. Moreover, the total biomass of onion in the experimental version was even slightly higher than in the control, mainly due to the more juicy and dense leaves.

The value of the integral TI index was 1.08, which indicated the comparative tolerance of onion plants to the presence in the $^{137}Cs$ ion environment, at least in the range of those short-term experiment conditions that were created in the current investigation.

**Table 1.** Average values and variability of biometric characteristics of onion under $^{137}Cs$ treatment and in control conditions.

| Index | $^{137}Cs$ treatment | Control |
|---|---|---|
| Shoot height (cm) | 13.0±3.5[a] | 16.6±1.2 |
| Root length (cm) | 7.9±2.6 | 6.8±2.3 |
| Fresh weight per 1 plant (g) | 4.5±0.30 | – |
| Leaves | 0.9±0.09 (20%) | – |
| Bulbs | 3.2±0.22 (71%) | – |
| Roots | 0.4±0.07 (9%) | – |
| Dry weight per 1 plant (g) | 1.42±0.07 | 1.16±0.06 |
| Leaves | 0.34±0.03 (24%) | 0.22±0.03 (19%) |
| Bulbs | 1.02±0.07 (72%) | 0.90±0.05 (78%) |
| Roots | 0.06±0.02 (4%) | 0.04±0.01 (3%) |

[a] hereafter confidence limits of the mean was calculated as $\pm t_{0.95} \cdot m$ at n = 7

*3.2. Visualization of $^{137}Cs$ root uptake and distribution in onion by digital autoradiography*

A digital autoradiographic study of the fractions of onion biomass grown on a $^{137}Cs$-containing medium showed the presence of a radionuclide in all organs of the plant (figures 2-4). Similar results for the transition of $^{137}Cs$ into crops of sunflower, reed and poplar under conditions of hydroponic experience were described by P. Soudek et al. [12, 13], who, based on a relatively uniform tonality of the image on autoradiographic prints, concluded that isotope was transported together with nutrients via the transpiration stream and distributed through this process across all tissues. This sharply distinguishes the behavior of $^{137}Cs$ in the "soil-plant" system from the behavior of a heavy natural radionuclide $^{238}U$, which is stored mainly in roots of a wide number of plant species cultivated under hydroponic conditions [14, 15, etc.].

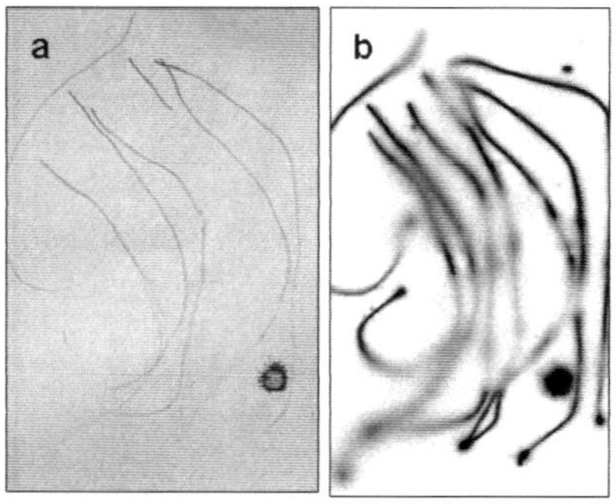


**Figure 2.** Accumulation of $^{137}Cs$ by the roots of onion: (a) photography, (b) autoradiography (expose time 48 hours).

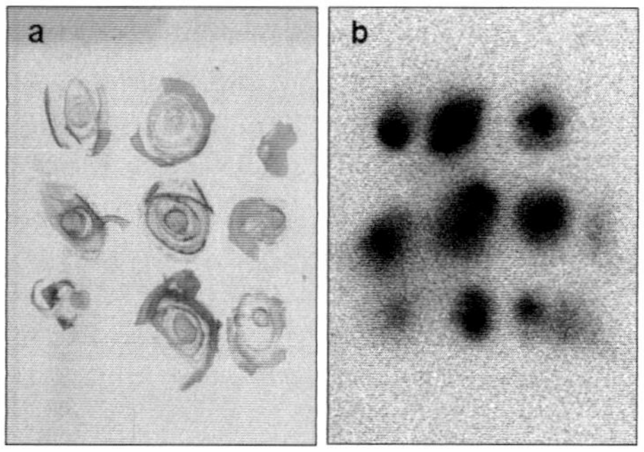


**Figure 3.** Accumulation of $^{137}Cs$ by the bulbs of onion: (a) photography, (b) autoradiography (expose time 48 hours).

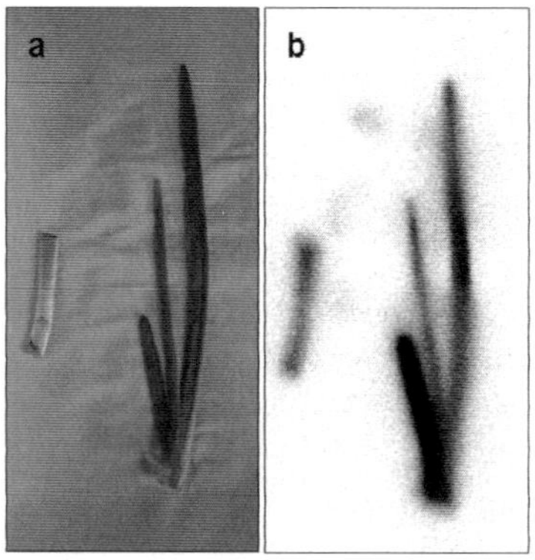


**Figure 4.** Accumulation of $^{137}Cs$ by the green leaves of onion: (a) photography, (b) autoradiography (expose time 48 hours).

By specifying the distribution of $^{137}Cs$ in plant organs by the tone of the image, the authors suggested that in plants the radionuclide was localized especially in young leaves, leaf veins and tips, in the nodes, young shoots, and in the root system as well. The instrumental resolution of the used in the current study autoradiograph device did not allow confirming or disproving such assumptions, it can only be asserted that the contrast of the plant's organ and its thickness and / or plant tissue density is obvious.

In particular, it is characteristic that the bottom of the bulb (flattened in the plate between the root and shoot parts of the plant) has a high contrast autoradiographic image, apparently not because of increased accumulation in this part of the radionuclide, but because of the increased thickness and density of the tissue. The same is seemingly valid for thickened green onion leaves or areas of their superposition on each other in a herbarium specimen. An important detail that goes beyond the observations is visualization of the process of preferential accumulation of $^{137}Cs$ on onion root tips, while over the rest of the length of the uniformly cylindrical root cross section, the radionuclide is distributed evenly. Since it is at the ends of the roots that the fission, stretching and absorption zones are concentrated, which are most actively involved in the growth and root nutrition of plants, then the increase in the accumulation of $^{137}Cs$ in them can have further cytogenetic and cytotoxic effects, leading to further chromosome aberrations. Thus, even in the absence of pronounced general symptoms of $^{137}Cs$ phytotoxicity in the hydroponic study, the possible negative damaging effect of radionuclide presence in the nutrient medium remains valid.

### *3.3. Assessment of $^{137}Cs$ root uptake and distribution in onion by γ-spectrometry*

Quantitative determination of $^{137}Cs$ in the biomass of onion and its fractions showed that the process of root consumption of the radionuclide was characterized by an average intensity: the specific activity of the radionuclide in plants was 3-4 orders of magnitude lower than in the hydroponic solution (table 2). Characteristically, despite the extremely high activity of the model solution on which onion was grown, as well as the availability of the ionic form of the radionuclide to plants, the intensity of $^{137}Cs$ transition into plants was less than the averaged predicted values of TF from soil to plant group of non-leafy vegetables: $3.5 \cdot 10^{-2}$ (sandy soils) – $9.1 \cdot 10^{-3}$ (clay soils) [8]. Thus, the ability to eliminate the transition of $^{137}Cs$ from nutrient media to plants is an inherent property of the latter, indicating the presence in plants of a number of metabolic and physiological adaptation mechanisms to cope with stress by decreasing toxic ions accumulation.

**Table 2.** Average values and variability of biometric characteristics of onion under $^{137}Cs$ treatment and in control conditions.

| Plant compartment | $^{137}Cs$ activity (Bq/g) | $^{137}Cs$ inventory per 1 plant (Bq) | % of $^{137}Cs$ inventory | TF / BC[a] | TC |
|---|---|---|---|---|---|
| Leaves | 424.2 | 144.2 | 24.5 | $0.15 \cdot 10^{-3}$ | 0.09 |
| Bulbs | 156.9 | 160.0 | 27.2 | $0.05 \cdot 10^{-3}$ | 0.03 |
| Roots | 4732.1 | 283.9 | 48.3 | $1.69 \cdot 10^{-3}$ | – |
| Total biomass | 414.2 | 588.2 | – | $0.24 \cdot 10^{-3}$ | – |

[a] TF value was calculated for total biomass, and BC was calculated for the biomass compartment, correspondingly.

At the same time, the highest values of $^{137}Cs$ activity, 10-30 times more than in the victorious part, were eliminated in the roots of onion. Significantly higher concentrations of $^{137}Cs$ in the roots suggest that the roots play in the plants in the cells. Taking into account the biomass structure of the plant, it can be argued that about $^{137}Cs$ deep, which passed from solution to plants, accumulated in the roots of onion, while the proportion of the roots themselves in the total biomass is less than 5% of the dry weight.

The intensity of $^{137}Cs$ from the roots of onion to non-chlorophyll flasks, as well as to green assimilating leaves, is 1–2 orders of magnitude smaller than the transition from solution to roots. TC values did not exceed 0.1, for both onion bulbs and leaves. At the same time, the specific activity of $^{137}Cs$ and the TC value in bulbs were significantly less than in the leaves. Perhaps, including the regulation of the specific plasma membrane of $^{137}Cs$ (excluding contact surface contamination) transporters.

It can be assumed that under conditions of $^{137}Cs$-stress, a strategy of toxic object rhizofiltration is implemented, in which the roots serve as a barrier that discriminates against the transition of the radionuclide to the escape. Radiocaesium fixed by roots, can protect more susceptible airborne parts from metal and radioactive toxicity [16] and oats [17], which, like the onion, belong to the class of monocotyledonous plants. An advantageous accumulation in the roots and a weak translocation into the aerial parts were also revealed with respect to a number of heavy metals known for their phytotoxic properties at extremely high concentrations of their mobile forms in soils – Cd, Pb, and Cu [10, 18–20]. Explaining the similar barrier function of the onion roots by the accumulation of the organic toxicant TNT (2,4,6-trinitrotoluene) grown as a hydroponic culture, Ja. Kim et al. [21] noted as possible causes for the heterogeneous distribution of matter between roots and green leaves: (1) the toxicant transported from the roots can be greatly diluted in the leaf volume, and (2) the translocation rate to the leaves from the roots is relatively low. Taking into account the last consideration, it can be assumed that for a longer experiment the $^{137}Cs$ distribution by the onion organs could change quantitatively, but it seems unlikely that a fundamental change in the structure of the radionuclide stocks will be possible.

Another assumption arose on the basis of the data obtained in the experiment: the value of the specific activity of $^{137}Cs$ and, as a consequence, the value of BC in the onion organ showed a directly proportional relationship with the relative difference between the wet and dry weight of the plant tissue, i.e. with the content of cellular juice in it (figure 5). Indeed, belonging to the group of monovalent alkali metals, $^{137}Cs$ is not capable of creating chelates or sparingly soluble salts, so it is quite logical to assume its occurrence in the intracellular fluid, rather than fixing the cell wall in the substance.

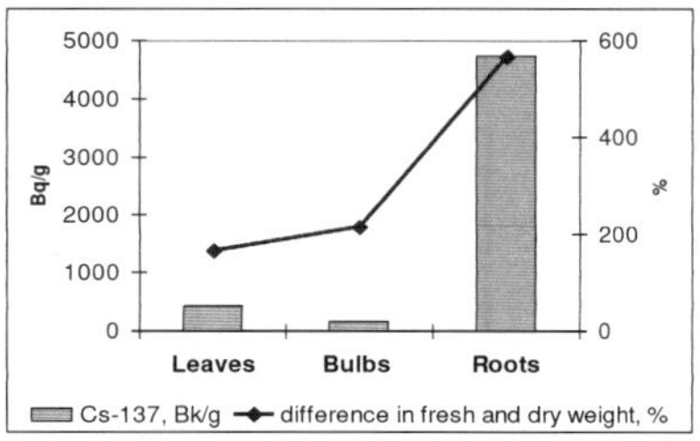


**Figure 5.** Relationship between $^{137}Cs$ activities in the onion compartment and the difference in fresh and dry weight of the correspondent compartment.

In general, during the model experience only a very small part of the radionuclide contained in the solution passed into plants. With the onion cultivation scheme used, the total PPF did not exceed 0.01%, and the total participation in the extraction of $^{137}Cs$ potentially easily removed from the soil of the shoot part of the plant did not exceed 0.006%. In this connection, the potential ability of onion plants to use them for phytoremediation of soils contaminated with $^{137}Cs$ should be considered to be of little promise.

## 4. Conclusions

Although hydroponic short-term studies do not reflect the real field of radioactively contaminated lands, it can be assumed that they are an efficient method for the express prediction of $^{137}Cs$ behaviour in the "soil-plant" system. Thus, in the present study it was shown that even under conditions of extremely high concentration of $^{137}Cs$ in the hydroponic solution, the transfer of the radionuclide to the onion was significantly limited due to the biological characteristics of the plant, which discriminate the process of root consumption of the toxicant. At the same time, biometric indices of plants – shoot height, roots elongation, biomass inventory and structure – under conditions of $^{137}Cs$ stress exposure were not significantly changed in comparison with the control. The preferential accumulation of $^{137}Cs$ occurred in the roots of the plant, and to the maximum extent, presumably at the root tips, most actively participating in the processes of mass exchange between the plant and the external environment. The translocation of $^{137}Cs$ from the roots to the shoot was significantly suppressed, which allows considering the root system of onion as a biological barrier that provides toxic filtration of the toxicant.

## Acknowledgments

The study was supported by the Russian Foundation for Basic Research (14-05-00903).

## References

[1] Alexakhin R M 2012 Radiation accidents – contribution to radioecological science and ecological lessons *Radioprotection* **44** 821–24

[2] 2015 *The Fukushima Daiichi accident* (Vienna: International Atomic Energy Agency) p 204

[3] Vodyanitskii Y N 2014 Natural and technogenic compounds of heavy metals in soils *Eurasian Soil Sci.* **47** 255-65

[4] Alexakhin R M, Buldakov L A, Gubanov V A, et al. 2004 *Large radiation accidents: consequences and protective countermeasures* (Moscow: IzdAT Publisher) pp. 330-86

[5] Puchkov V A and Bolshov L A (Eds) 2016 *Russian National Report 30 Years of the Chernobyl Accident: Results and Prospects for Overcoming its Consequences in Russia. 1986-2016* (Moscow: Ministry of the Russian Federation for civil defense, emergency situations and Elimination of Consequences of Natural Disasters) pp. 113-37 (*In Russian*)

[6] Fesenko S V, Alexakhin R M and Balonov M I 2007 An extended critical review of twenty years of countermeasures used in agriculture after the Chernobyl accident. *Sci. Total Environ.* **383** 1–24

[7] Paramonova T, Belyaev V, Komissarova O and Ivanov M 2017 Homo/heterogeneity of Cs-137 distribution within ploughed horizon of arable chernozems, 30 years after Chernobyl accident *Radiation and Applications* **2** 192–9

[8] 2010 *Handbook of Parameter Values for the Prediction of Radionuclide Transfer in Terrestrial and Freshwater Environments* (Vienna: International Atomic Energy Agency) p 194

[9] Zacchini M, Pietrini F, Mugnozza G S, Iori V, Pietrosanti L and Massacci A 2009 Metal tolerance, accumulation and translocation in poplar and willow clones treated with cadmium in hydroponics *Water Air Soil Pollut.* **197** 23–34

[10] Juang K-W, Lee Y-I, Lai H-Yu, Wang Ch-H and Chen B-Ch 2012 Copper accumulation, translocation, and toxic effects in grapevine cuttings *Environ. Sci. Pollut. Res.* **19** 1315–22

[11] Oudalova A A, Geras'kin S A, Dikareva N S and Pyatkova S V 2017 Allium-test as a tool for toxicity testing of environmental radioactive-chemical mixtures *J. Phys.: Conf. Ser.* 784 012057

[12] Soudek P, Tykva R and Vanek T 2004 Laboratory analyses of $^{137}$Cs uptake by sunflower, reed and poplar *Chemosphere* **55** 1081–7

[13] Soudek P, Valenova S, Vavrıkova Z and Vanek T 2006 $^{137}$Cs and $^{90}$Sr uptake by sunflower cultivated under hydroponic conditions *J. Environ. Radioact.* **88** 236–50

[14] Straczek A, Duquene L, Wegrzynek D, China-Cano E, Wannijn J, Navez J and Vandenhove H 2010 Differences in U root-to-shoot translocation between plant species explained by U distribution in roots *J. Environ. Radioact.* **101** 258–66

[15] Soudek P, Petrova S, Benesova D, Dvorakova M and Vanek T 2011 Uranium uptake by hydroponically cultivated crop plants *J. Environ. Radioact.* **102** 598–604

[16] Shaw G, Hewamanna R, Lillywhite J and Bell J N B 1992 Radiocaesium uptake and translocation in wheat with reference to the transfer factor concept and ion competition effects *J. Envir. Radioact.* **16** 167–80

[17] Paramonova T A, Kuzmenkova N V, Godyaeva M M, Belyaev V R, Ivanov M M and Agapkina G I 2018 Cesium-137 root uptake by oat and lettuce test crops from radioactively contaminated chernozem under model experiment conditions *Moscow Univ. Soil Sci. Bullet.* **73** 18–25

[18] Siebers N, Siangliw M and Tongcumpou Ch. 2013 Cadmium uptake and subcellular distribution in rice plants as affected by phosphorus: Soil and hydroponic experiments *J. Soil Sci. Plant Nutrition* **13** 833-44

[19] Chen Z, Tang Y-T, Yao A-J, Cao J, Wu Z-H, Peng Z-R, Wang S-Z, Xiao S, Baker A and Qiu R-L 2017 Mitigation of Cd accumulation in paddy rice (*Oryza sativa* L.) by Fe fertilization *Environment. Pollut.* **231** 549–59

[20] Koubova K, Tlustos P, Brendova K, Szakova J and Najmanova J 2016 Lead accumulation ability of selected plants of Noccaea spp *Soil Sediment Contamination* **25** 882–90

[21] Kim Ja, Drew M C and Corapcioglu M Ya 2004 Uptake and phytotoxicity of TNT in onion plant *J. Environ. Sci. and Health. Part A – Toxic/Hazardous Substances & Environmental Engineering* **A39** 803–19

# Comparison of DNA damage level in irradiated blood cells of mouse, frog and human

**I Y Mitroshina, N P Sirota and E A Kuznetsova**

Institute of Theoretical and Experimental Biophysics of the Russian Academy of Science, Moscow Region, Institutskaya st. 3, 142290 Pushchino Russia

E-mail: xf2@rambler.ru

**Abstract**. Here, we present the comparative studies of DNA damage induced by X-ray irradiation in vitro in mouse, frog and human blood cells and assessed by the "comet test" (alkaline version). We showed that, in contrast to mouse and human cells, DNA repair processes in frog blood cells are not completed by 60 minutes of the post-radiation period. The results suggest that frogs can be used as indicators of environmental radiation pollution. In this regard, additional experiments are needed to assess the different types of external and internal exposure of animals.

## 1. Introduction

Studying the effects of ionizing radiation (IR) on living organisms remains important, because there is always a possibility of radiation exposure when using radioactive materials for industrial and medical purposes as well as during radiation accidents. Fast and minimally invasive methods are required for an individual retrospective assessment of the radiation response in the body. The assessment is carried out mainly using hematological and cytogenetic methods. The analysis of chromosome aberrations remains the "gold standard" for early biodosimetry in case of accidents, but it is a time-consuming procedure [1]. Therefore, the study of the processes occurring in cells and tissues of living organisms after irradiation including the evaluation of DNA damage levels and the detection of sensitive objects of radiation exposure are in demand at the present time.

To determine the levels of DNA damage in individual cells from tissues of various organisms, including human leukocytes, the method "Comet test" is used [2, 3]. The alkaline version of the method makes it possible to reveal a wide range of DNA damage: single- and double-strand breaks, alkaline-labile sites and sites with damaged bases [4].

To assess environmental changes, comparative studies of DNA damage levels in cells obtained from organisms living in contaminated and background areas are needed. Currently, the whole blood has become one of the most useful objects for assessing the level of DNA damage by different methods, including the "Comet test" method. Blood in members of different taxons differs in blood cell types, and it is not known how this difference correlates with the levels of DNA damage in blood cells.

In this work, we carried out comparative studies of DNA damage levels in blood cells mice, frogs and humans after radiation exposure in vitro. Frog blood was taken for comparison with mouse and human blood, as frogs are considered as sensitive indicators for environmental pollution [5]. The level of DNA damage in the frog's blood cells was evaluated both immediately and 1 hour after irradiation.

## 2. Materials and Methods

*2.1 Cells*

In the course of the work, 10–15 μl of capillary blood of healthy donors aged 25–65 years (12 men, 8 women) were used. At the time of blood sampling, donors did not exacerbate infectious or chronic diseases. The human blood was selected for evacuators containing K-EDTA. Informed consent was obtained from each participant before the blood draw.

The blood of the frog was kindly presented by V.K. Uteshev. Whole blood samples of 15 male *Rana temporaria* aged 2–3 years used. The blood from decapitated frogs, 50 μl from each, was collected in the tube with 0.3 ml of cold PBS, 1 mM EDTA.

Mice blood was obtained by incising the tip of the tail of 20 male animals with a mass of 22–25 g. The noninbred SHK strain mice were obtained from the vivarium of the ITEB RAS. Blood samples were collected in tubes with phosphate buffered saline (PBS), pH 7.2, containing 1 mmol/L EDTA (PBS-EDTA). Splenocytes from mice were obtained according to [6]. The experiments were carried out according to the requirements set by the Institute Committee on Control for the use of animals in biomedical experiments.

Whole blood of mice and human was used, diluted with PBS-EDTA 6 times; frog blood and splenocyte suspensions were diluted with PBS-EDTA to $10^6$ cells/ml.

*2.2 Irradiation*

Cells / nucleoids were exposed to X-ray radiation within agarose slides in RTD-250-15-1 X-ray unit (Russia) operated at 200 kV and of 20 mA. Dose rate was 1.12 Gy/min. Radiation was filtered through 1 mm Al and 1 mm Cu. The focal length was 37 cm. Immediately after irradiation, the slides were placed in a cooled lysis solution.

Part of the slides containing the frog erythrocytes after irradiation was placed in PBS or medium 199 (Gibco Lab). Next, the slides were incubated for 60 minutes at 37 °C.

*2.3 Comet Assay*

First, slides were placed in a solution of 1% agarose and dried. A 1% agarose layer was applied to these glasses and incubated in the refrigerator until it solidified (5–7 minutes). 1% low- melting agarose (Sigma Chem Co, USA) was melted at 70 °C in PBS-EDTA and incubated to 20–22 °C. The blood or cell suspension was mixed with an equal volume of the 1% low-melting agarose. 15 μl of this mixture was applied to the prepared agarose layer. The agarose layer with the cells was cooled until it solidified. A new layer of 0.5% low-melting agarose was applied over the cooled layer with the cells. The slides were genotoxically exposed and placed in a lysis solution (2.5 mol / L NaCl, 0.1 mol / l EDTA, 0.01 mol/L Tris-HCl, pH 10, 1% Triton X-100) at 4 °C for 1 hour.Then the slides were exposed for 20 minutes in alkaline solution "A" (0.3 mol/L NaOH, 0.001 mol / l EDTA, pH> 13). After alkaline denaturation, the slides were moved to an electrophoretic chamber SE-1 / S-1N (OOO Helikon, Russia) and electrophoresis was carry out in a fresh portion of solution "A" for 20 minutes at 4 °C. The buffer volume was 250 ml, voltage 27 V, amperage 260-270 mA (electric field strength 2 V / cm). After electrophoresis, the slides were washed with distilled water and stained with ethidium bromide (2.0 μg / ml) for 1 hour. The preparations were analyzed using a fluorescent microscope LUMAM I-3 (LOMO, St. Petersburg, Russia). The image capturing was carried out by a photocamera Nikon CoolPix 995 (Japan). Received images transferred to the computer. The analysis of images was made by means of special software developed to register and analyze the images, where algorithms of computation of the standard comet parameters have been realized [7].

The level of DNA lesions was judged from the fraction of DNA in comet tails in percent – % TDNA. For each experimental point, at least 5 individual animals were taken, 3 slides for each one, 50–75 comets were scored per slide, according to recommendations [8]. The statistical analysis was performed by Student's t-test ($p < 0.05$).

## 3. Results and Discussion

At the initial stage of the study, preliminary experiments were carried out in which the levels of DNA damage in irradiated cells and in nucleoids obtained from these cells were compared. Agarose slides containing mouse splenocytes were prepared, some of which underwent a lysis procedure for the preparation of nucleoids. After that, slides containing both whole cells and nucleoids were irradiated at 2, 4, 8 Gy, and the levels of DNA damage were measured. Figure 1 shows correlation between the level of DNA damage and the dose of X-ray irradiation in splenocyte nucleoids and whole splenocytes from a mouse irradiated at room temperature.

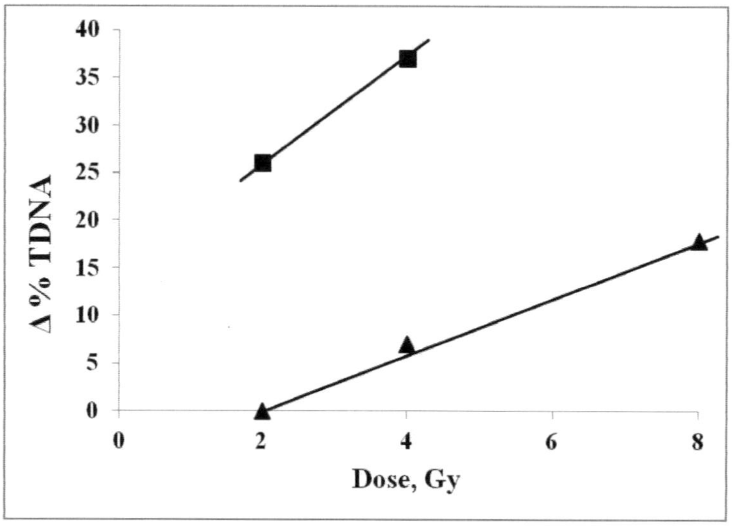


**Figure 1.** Dependence of the DNA damage level of mouse splenocyte nucleoids (■) and whole splenocytes (▲) on the irradiation dose.

The damage level is shown as the difference in % TDNA values between exposed and control samples. This figure shows that at the same dose, % TDNA in nucleoids is higher compared to splenocytes, what indicates active DNA repair in splenocytes. The possibility of DNA repair in the blood cells within the agarose slides was evaluated by incubating the slides in the PBS for 1 hour at 37 °C.

Figure 2 shows the change in % TDNA in human, mouse and frog blood cells after irradiation of these cells at a dose of 4 Gy at room temperature and subsequent incubation in the PBS at 37 °C for 60 minutes.

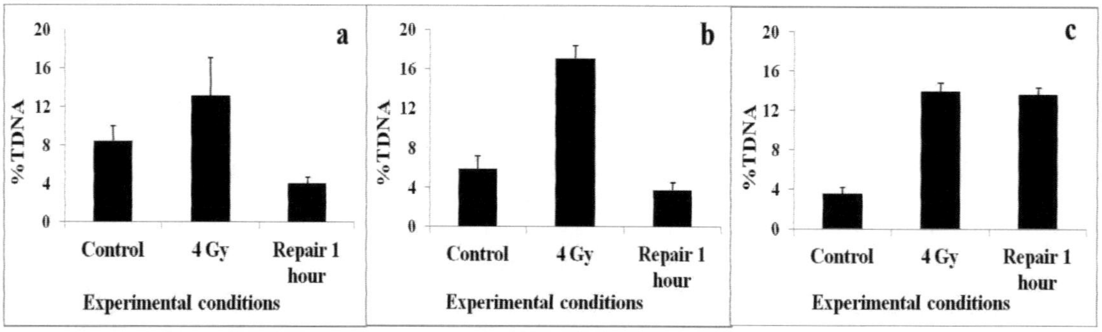


**Figure 2.** The level of DNA damages of nucleated cells of peripheral blood of a human (a), mouse (b) and frog (c) after irradiation at a dose of 4 Gy and incubation for 60 minutes (M ± SEM)

It can be seen that the level of DNA damage in control cells of a human, mouse and frog does not exceed 10%. This basic level is due to both: the functioning / metabolism of DNA and the continuous emergence of DNA lesions, including oxidative modifications, and their repair [9]. After exposure to X-rays at a dose of 4 Gy %TDNA increased in the cells of all organisms and it differed from the basic level significantly. After 60 minutes incubation, it significantly decreased only in mouse and human leukocytes, reaching a control level, what is consistent with the literature data [10, 11]. In frog blood cells 60 minutes after irradiation,% TDNA was comparable to that one measured immediately after irradiation, what indicates a very slow DNA repair.

In other series of experiments, the frog blood cells was irradiated at a dose of 8 Gy and incubated for 60 minutes at 37 °C in both PBS and medium 199 (figure 3).

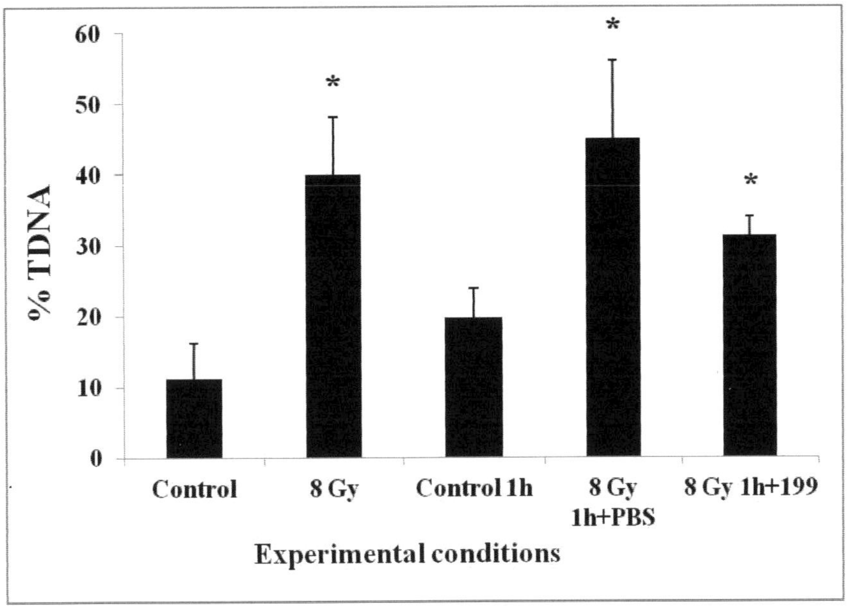


**Figure 3.** The level of DNA damage of the frog's blood cells after irradiating at a dose of 8 Gy at room temperature (M ± SD). 1h – 1 hour. 199 – medium 199. * – significant difference from t-test ($p < 0.05$)

% TDNA in blood cells irradiated at a dose of 8 Gy was significantly higher than at 4 Gy and after 60 minutes incubation in the PBS did not differ from level measured immediately after irradiation. After incubation in the medium 199, % TDNA decreased compared to that one in PBS, but no significant difference was found.

Note that the blood of a mouse, a human and a frog differs both in blood cell types and in their ratio. Lymphocytes form the largest group among the white blood cells in mice and neutrophils – in humans. Among the frog blood cells, nucleated red blood cells predominate. Moreover, metabolic processes in amphibians are not as active as in mammals, and it can be expected that the processes of DNA repair also proceed more slowly than in mammals. Thus, the repair of radiation-induced DNA sites sensitive to formamidopyrimidine-glycosylase was more rapid in rodent cells than in mammalian cells. It suggested the higher level of DNA polymerase β activity [10].

Unlike mammals, the nucleated erythrocytes of amphibians and birds additionally contain histone H5, which appears to play an important role in maintaining the repressed state of chromatin [12, 13]. For example, it was shown that in the nucleated erythrocytes of chickens there was no repair of single-stranded DNA breaks induced by ionizing radiation [14].

## 4. Conclusions

We compared the levels of DNA damage in blood cells in members of different taxons.

It can be assumed that the level of DNA damage depends on the blood cells composition and their metabolism but not their radiosensitivity.

DNA repair processes in frog blood cells are not completed by 60 minutes of the post-radiation period and the level of DNA damage remains high. Thus, frogs can be used as indicators of radiation pollution of the environment both in the background and in emergency situations.

Further studies should be performed to estimate of different types of external and internal irradiation of animals.

**Acknowledgments**

We are grateful to V.K. Uteshev (Institute of Cell Biophysics RAS, Pushchino of Moscow Region) for granting blood samples of Rana temporaria.

## References

[1] Wang Y, Xu C, Du L Q, Cao J, Liu J X, Su X, Zhao H, Fan F-Y, Wang B, Katsube T et al 2013 Evolution of the Comet Assay for assesing the dose-response relationship of DNA damage induced by ionizing radiation *Int J Mol Sci* **14** 22449–61

[2] Graupner A, Eide D M, Instanes C, Andersen J M, Brede D A, Dertinger S D, Lind O C Brandt-Kjelsen A, Bjerke H, Salbu B et al 2016 Gamma radiation at a human relevant low dose rate is genotoxic in mice *Sci. Rep.* **6** 32977

[3] Møller P 2018 The comet assay: ready for 30 more years *Mutagenesis* **33** 1–7

[4] Collins A R, Oscoz A A, Brunborg G, Gaivão I, Giovannelli L, Kruszewski M, Smith C C and Stetina R 2008 The comet assay: topical issues *Mutagenesis* **23** 143–51

[5] Cohen M M Jr 2001 Frog decline, frog malformations, and a comparison of frog and human health *Am J Med Genet* **104** 101–9

[6] Kondratieva I 2001 *Manual in Immunology* eds I Kondratieva and V Samuilov (Moscow: Moscow State University Press) pp 8–41

[7] Chemeris N K, Gapeyev A B and Sirota N P 2004 The in vitro assessment of potential genotoxicity of high power microwave pulses *Mutat Res* **558** 27–34

[8] Lovell D P and Omori T 2008 Statistical issues in the use of the comet assay *Mutagenesis* **23** 171–82

[9] Davalli P, Marverti G, Lauriola A and D'Arca D 2018 Targeting oxidatively induced DNA damage response in cancer: opportunities for novel cancer therapies *Oxid Med Cell Longev* **2018** 2389523

[10] Purschke M, Kasten-Pisula U, Brammer I and Dikomey E 2004 Human and rodent cell lines showing no differences in the induction but differing in the repair kinetics of radiation-induced DNA base damage *Int J Radiat Biol* **80** 29–38

[11] Saha D T, Davidson B J, Wang A, Pollock A J, Orden R A and Goldman R 2008 Quantification of DNA repair capacity in whole blood of patients with head and neck cancer and healthy donors by comet assay *Mutat Res* **650** 55–62

[12] Bradbury E M, Molgaard H_V, Stephens R_M, Bolund L_A and Johns E W 1972 X-ray studies of nucleoproteins depleted of lysine-rich histone *Eur J Biochem* **31** 474–82

[13] Sung M T, Harford J, Bundman M and Vidalakas G 1977 Metabolism of histones in avian erythroid cells *Biochemistry* **16** 279–85

[14] Karran P and Ormerod M G 1973 Is the ability to repair damage to DNA related to the proliferative capacity of a cell? The rejoining of X-ray-produced strand breaks *Biochim Biophys Acta* **299** 54–64

IOP Conf. Series: Materials Science and Engineering **487** (2019) 012017 doi:10.1088/1757-899X/487/1/012017

# Bottom sediments as indicators of water resources state within the town boundaries of Cherepovets, Vologda region

**IA Neporozhnaya, EL Petrov and DN Kalinina**

Cherepovets State University, Lunocharsky pr. 5, Cherepovets, Russia

E-mail: nep-inna@yandex.ru

**Abstract.** The article considers the research of the plant pigments content in the upper layer of the bottom sediments in the rivers (the Serovka, the Yagorba, the Sheksna), which are affected by communal and industrial wastewater in Cherepovets. The average content of pigments in the bottom sediments of the Serovka and the Yagorba rivers was higher than in the Sheksna river during the observation period (July 2015 – December 2016). According to the indices E480 / E665, E480 / (1.7E665k) and the ratio of chlorophyll to pheophytin, the pigmentary fund of benthic algocenoses is mainly in a degraded state. In terms of chlorophyll and pheopigments (calculated on dry soil) amount, the bottom sediments of the Serovka river are characterized by the mesotrophic category values, while the ones of the Yagorba and the Sheksna rivers – by oligotrophic values.

## 1. Introduction

The rivers flowing in the town are used to provide drinking water, irrigation of agricultural land, industrial and municipal water supply and wastewater discharge, navigation, fishing and recreation. River water quality largely depends on the bottom sediments state. Annually a huge amount of industrial and communal wastewater is discharged into reservoirs. It leads to the accumulation of various substances in the bottom sediments, and affects the processes in the reservoirs. Sewage can lead to eutrophication of water bodies. At present water bodies eutrophication is global in scope. The main indicator of eutrophication is primary production. To determine water bodies productivity various indicators are used. They include the content of chlorophyll, pheophytin, carotenoids, and pigment indices [1–4]. Bottom sediments are considered the best indicators of eutrophication, since they are sources of internal nutrient loading, which largely determines reservoir productivity [5–7]. Determination of the plant pigments content in the bottom sediments makes it possible to assess the ecological state of water bodies, and to predict their ecological state for the future.

The research is aimed at estimating of the rivers ecological state within the boundaries of Cherepovets, Vologda region, on the basis of pigment characteristics of the bottom sediments. Cherepovets is the largest town in Vologda region, the administrative center of Cherepovets district. It is located on the Sheksna river, on the left tributary of the Volga, at the mouth of the Yagorba river. The Sheksna, the Yagorba, the Serovka, the Nelaz, the Torovka, and the Koshta rivers flow in the town and its surrounding areas [8]. Characteristic feature of all the rivers in the Cherepovets area is their regulation by a backwater from the Rybinsk reservoir, resulting in the formation of deep-water bays in the mouths of the rivers. The entire water area of the northern part of the Sheksna and the small rivers of this region are under the powerful influence of Cherepovets. It receives heated urban water, a large number of organic compounds of domestic and industrial origin. Urban sewage spreads over a distance of more than 11 kilometers along the Sheksna River, and the effect of sewage in the plume manifests itself at a distance up to 25-55 kilometers. Severe contamination with phenols, naphthalene derivatives, petroleum products, fecal discharges leads to a sharp increase in the number of bacteria. The Koshta, the Sheksna, the Serovka rivers, the mouth of the Yagorba river are severely polluted. The Sheksna river undercurrent has a significant influence on the water formation of the northern part of the Sheksna reach in the Rybinsk Reservoir, and is strongly affected by the tributaries and upper areas of the Sheksna reservoir [9]. The greatest anthropogenic load is on the water objects, on the banks of which there are large industrial hubs, including Cherepovets (the Koshta, the Yagorba, the Serovka rivers) [10]. On the banks of all these rivers there are industrial enterprises: JSC "Cherepovets Port", JSC "Severstal-Invest", CJSC "Cherepovets Plywood and Furniture Factory", the river berths for the country passenger ships, JSC "Cherepovets Match Factory FESKO", JSC "Cherepovets Shipbuilding and Shiprepairing Plant", cattle-breeding complex (CJSC "Botovo") and summer cottages.

## 2. Methods

The research samples were selected on the territory of Cherepovets town from July 2015 to December 2016 in the Sheksna, the Yagorba and the Serovka rivers by the GR91 bottom-grab from the upper layer of the bottom sediments (0–5 cm) in the shallow areas, two meters from the water edge (at the depth of 1–1.5 m), one time per month during the year. Plant pigments in the bottom sediments were determined by spectrophotometric method on the PE-5400uf spectrophotometer in acetone extract, the calculations were carried out according to the Lorentz equations, generally accepted for phytoplankton, with some modifications [11, 12]. The indicator of the pigmentary stock degradation degree (the ratio of carotenoids and chlorophyll concentrations) was estimated from the extract optical densities at wavelengths of 480, 665 nm before and after acidification by the E480 / E665 and E480 / 1.7 * E665k indices [11]. The soil moisture content and the amount of organic substance were determined by traditional methods [13]. The river trophic state estimate was computed on the basis of the sum of chlorophyll and phenopigments calculated on dry soil: <13 μg / g oligotrophic, 13–60 μg/g mesotrophic, 60–120 μg / g eutrophic, > 120 μg/g hypertrophic [14].

## 3. Results and Discussion

The bottom sediments of the rivers under study differed significantly in the content of photosynthetic pigments, organic substance and humidity. The chlorophyll content in the bottom sediments of the Serov river ranges from 0.83 to 29.5 μg/g, pheophytin – from 0.78 to 34.4 μg/g, carotenoids – from to 8.7 to 48.4 μg/g. The chlorophyll content in the bottom sediments of the Yagorby river ranges from 0.3 to 20.9 μg/g, pheophytin – from 0.1 to 31.8 μg/g, carotenoids – from 1.5 to 46.6 μg/g. The chlorophyll content in the bottom sediments of the Sheksna River fluctuates from 0.21 to 5.5 μg/g, pheophytin from 0.29 to 7.2 μg/g, carotenoids from 1.5 to 9.3 μg/g. The content of carotenoids in the of river bottom sediments considerably exceeds the content of chlorophyll, which may be a consequence of stratification leading to changes in destructive processes [11], different rates of plant pigments destruction: carotenoids are destroyed more slowly. The average content of chlorophyll with pheophytin (Ch + F) in the upper layer of the Serovka (17.8±3.1 μg/g) and the Yagorba (12.9±2.9 μg/g) bottom sediments was characterized by higher values than in the Sheksna river 5.0±0.55 μg/g). The greater pigment content over the period under study was observed in the Serovka river (figure 1) than in the Yagorba and the Sheksna rivers. The intake of phytoplankton organic substance in the bottom sediments depends on the flow speed and on the anthropogenic influence – the inputs of nutrients due to sewage from the industrial enterprises and from local settlements situated on the banks of the rivers. The River Sekovka differs from the Yagorba and the Sheksna rivers by a lower flow rate. It increases the nutrient content in the water and, in combination with the weak hydrodynamic regime, affects the siltation of the reservoir.

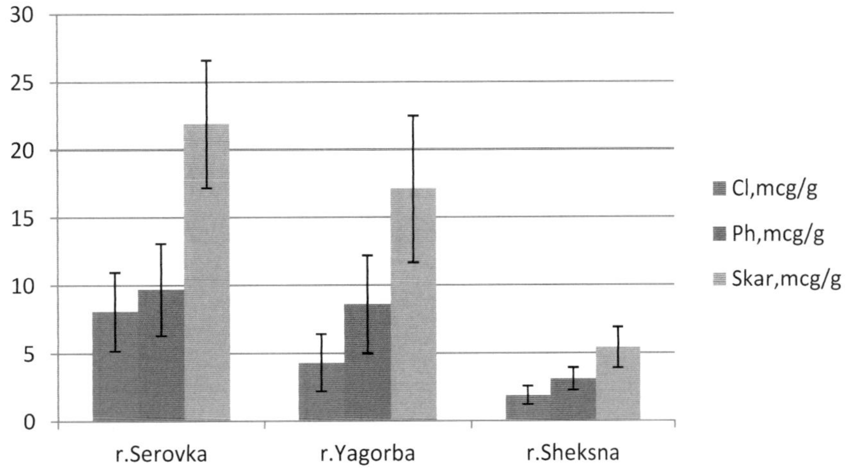


**Figure 1.** The average content of pigments in the bottom sediments of the Cherepovets rivers (from July 2015 to December 2016).

By soil type: the Serovka river – gray silt; the Yagorba river – silty sand; the Sheksna – sand. The content of water (humidity) in soils is directly dependent on the soil type. Large and medium-grained sand is characterized by low humidity (11–24%). The highest moisture content is typical for silts (15–55%). Silty sands keep an intermediate position (19–31%).

The average pigment content clearly increased in the series: sand (the Sheksna river) → silty sand (the Yagorba river) → gray silt (the Serovka river) (figure 2). According to the analysis of the plant pigments content, organic substance and moisture, the investigated rivers can be arranged in the following order: the Serrovka river > the Yagorba river > the Sheksna river.

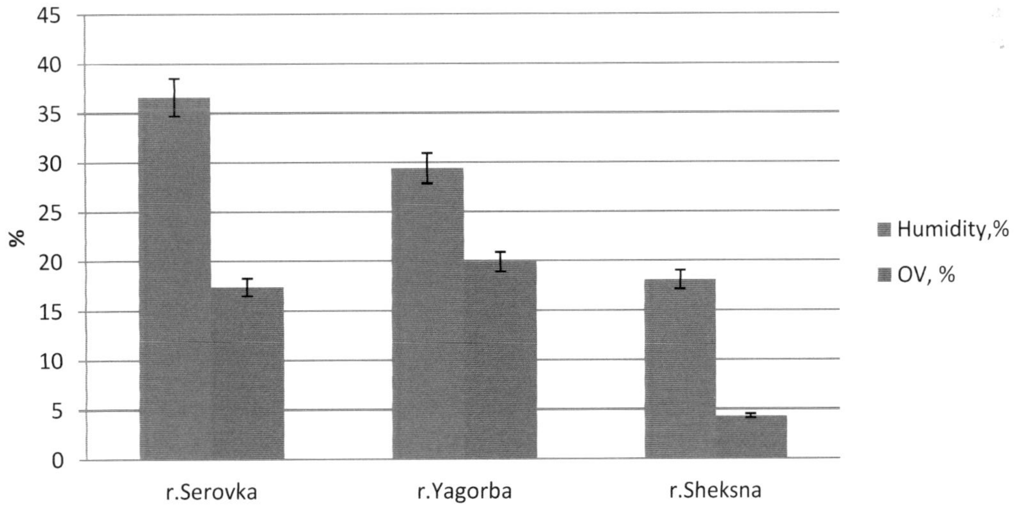


**Figure 2.** Humidity and organic substance (OV) content in the bottom sediments of the Cherepovets rivers (from July 2015 to December 2016).

The plant pigments ratio in the water and bottom sediments reflects the result of the synthesis and destruction of the newly formed organic substance, the result of the photosynthetic apparatus of algae and their pigments transformation [15] (Table 1). Chlorophyll in the river bottom sediments is represented mainly by the products of destruction – from 60 to 98% of their total amount with chlorophyll.

**Table 1.** The ratio of plant pigments in the bottom sediments of Cherepovets rivers.

| N | Pigmentary indices | The Serovka river | The Yagorba river | The Sheksna river |
|---|---|---|---|---|
| 1 | 2 | 3.00 | 4.00 | 5.00 |
| 1 | $E_{480}/E_{665}$ | 3.08 | 2.60 | 2.60 |
| 2 | $E_{480}/(1.7E_{665K})$ | 2.40 | 1.69 | 1.61 |
| 3 | Chl/Ph | 0.59 | 0.49 | 0.55 |

In the river bottom sediments, the index E480 / E665 changed in the Serovka river from 1.17 to 9.33, in the Yagorba river from 1.07 to 4.58, and in the Sheksna river – from 1.04 to 3.83 at different observation time. Clear seasonal dynamics of the index values were not revealed. The average values of the yellow and green pigments ratios (E480 / E665) (Table 1) exceed the values for the functioning phytoplankton. The correction for the presence of chlorophyll derivatives leads to a slight decrease in the concentration ratio of yellow and green pigments (E480 / (1.7E665k)) (Table 1), but this index does not reach the values characteristic of phytoplankton. Perhaps this is due to the predominant degradation and discoloration of chlorophyll, the accumulation of transformed organic substance. The ratio of chlorophyll pigments to pheophytin (Chl / Ph) is less than one, indicating the extinction and decomposition of algae [16–18]. The obtained physiological state indices determine the ecological state of phytoplankton as altered, functionally impaired. The rivers under study experience long-term impact of communal and industrial wastewater, as a result of which, in excess amount, various compounds enter the water of these rivers, which adversely affects the structure and functioning of plankton and benthic communities [19], as well as the bottom sediments composition. The qualitative composition of the bottom sediments is the most important factor of the environmental state, since it determines the amount of pollutants deposition, the scale and rate of their entry into the water masses and biota [20].

According to the results of the study, it is possible to estimate the trophic state of the rivers by sedimentary pigments: the Serovka river can be referred to mesotrophic, the Jagorba and the Sheksna rivers – to oligotrophic reservoirs. Of the rivers under investigation, the Serovka river is the most productive in comparison with the Sheksna and the Yagorba rivers, but in all these rivers destructive processes predominate, which determines the ecological state of water bodies as altered, functionally impaired.

## References

[1] OECD 1982 *Eutrophication of waters: Monitoring, assessment and control* (Paris: Organization of Economic Cooperation and Development) p 155

[2] Devyatkin V G, Meteleva N Yu and Mitropolskaya I V 2001 Hydrophysical factors of prodctivity of littoral phytoplankton: estimation and prognosis of the content of chlorophyll a and intensity of photosynthesis *Biol. int. water* **1** 36–45

[3] Pyryn I L, Sigareva L E 2005 Phytoplankton pigment content as an indicator of the current state of the Rybinsk reservoir *Actual problems of ecology in the Yaroslavl Region issue 3 Mat. Conf.* V 1 (Yaroslavl) pp 270–4

[4] Pyryna I L, Litvinov A S, Kuchai L A, Roshchupko V F and Sokolova E N 2006 Long-term changes in the primary production of phytoplankton of the Rybinsk Reservoir in connection with the effect of climatic factors *The state and problems of production hydrobiology*

(Moscow: KMC) pp 36-46
[5] Rossolimo L L 1977 Changes in liminal ecosystems under the influence of anthropogenic factors *Science* (Moscow) p 144
[6] Datsenko Yu S 2007 Eutrophication of reservoirs. Hydrological and hydrochemical aspects *GEOS* (Moscow) p 252
[7] Sigareva L Y and Timofeeva N A 2011 Interannual dynamics of chlorophyll a content in bottom sediments of the Rybinsk Reservoir (Russia) *Algology* **21** 190–201
[8] Vorobyev G A 2007 Nature of the Vologda region (Vologda: Vologzhanin Publishing House) p 440
[9] Bolotova N L 1996 The Edge of the Man-Made Sea (Water Resources of the Cherepovets District) *Cherepovets: Local Lore. al. / City Hall of Cherepovets, Cherepovets. museums* (Vologda. ped. un-t. - Vologda: Russia) V 3 p 318
[10] 2017 *Report on the State and Protection of the Environment in the Vologda Region in 2016* (Vologda: Government of the Vologda region, Department of Natural Resources and Environmental Protection of the Vologda region) p 250
[11] Sigareva L Y 2012 Chlorophyll in the bottom sediments of the Volga reservoirs *Association of Scientific Publications KMK* (Moscow) p 217
[12] Merkushina G A and Larina N S 2013 Peculiarities of spectrophotometric determination of photosynthetic pigments in peat of different genesis *Pros. of KazNU. Chemical Series* **4** 103–10
[13] Butorin N V, Ziminova N A and Kurdin V P 1975 Bottom sediments of Upper Volga reservoirs *Science* (Leningrad) p 159
[14] Möller W A A and Scharf B W 1986 The content of chlorophyll in the sediment of the volcanic maar lakes in the Eifel region (Germany) as an indicator for eutrophication *Hydrobiologia* **143** 327–9
[15] Takhtadzhyan A L 1977 *Life of plants: in 6 volumes V 3 Algae. Lichens (*Moscow: Prosveschenie) p 487
[16] Britton G. 1986 *Biochemistry of Natural Pigments* (Moscow: Mir) p 425
[17] Mineeva N. 2004 Vegetable pigments in the water of the Volga reservoirs *Science* (Moscow) p 156
[18] Belaya S A and Khristoforova N K 2011 Photosynthetic pigments of periphyton algae in the waterways of the Sikhote-Alin Biosphere Reserve *Reading memories of V.Ya. Levanidova* Issue 5 pp 53–60
[19] Kopylov A I, Ievleva T V, Romanenko A V and Zabotkina EA 2015 Structural and functional characteristics of bacterioplankton of rivers flowing through a large city (Cherepovets, Verkhnaya Volga) *Povolzhsky Ecological Journal* **2** 183–92
[20] Nazarov N N and Suntsov A V 2008 Geochemical horizons and microelement composition of bottom sediments of the Votkinsky reservoir *Geografichesky vestnik* **2** 22–6

# The effects of electromagnetic field and γ-radiation on cell proliferation and regeneration in planarian *Schmidtea mediterranea*

**E I Sarapultseva[1,2,3], O N Matchuk[2], D Uskalova[1], H Markina[1], N Savina[1] and I A Zamulaeva[1,2]**

[1] Institute of Nuclear Power Engineering NRNU MEPhI, Obninsk, Russia
[2] A. Tsyb Medical Radiological Research Center, National Medical Research Radiological Center, Ministry of Health of the Russian Federation, Obninsk, Russia
[3] National Research Nuclear University MEPhI, Moscow, Russia

Email: helen-bio@yandex.ru

**Abstract**. We have shown that γ-irradiation at dose of 10 Gy and low-intensity electromagnetic fields (EMF, 900 MHz, 100 μW/cm$^2$, 180 min) significantly compromised cell proliferation of planarian *S. mediterranea* regenerating after decapitation. Using flow cytometry we have shown that 24 hours after exposure to γ-rays and EMF the index of proliferation established as the total amount of cells at S and G2/M phases was 2.8 and 1.8 lower than that in controls respectively. Our data therefore provide the first experimental evidence for the effects of medium-dose γ-irradiation on regeneration in planaria which is attributed to radiation-induced compromised proliferation activity.

## 1. Introduction

In recent years one of the most important problems of radiobiology and radioecology is considered to clarify radiation effects on biology of stem cells that play a key role in maintaining tissue homeostasis during adult life and tissue repair processes after damage. It is well known that neoblasts being stem cells make up 30% of the total number of cells in the multicellular organism free-living fresh-water planarian flatworms [1–3]. Planarians are characterized by a unique ability to regenerate lost or damaged body parts and provides an excellent opportunity for the study of post-traumatic recovery processes. Currently, planarians are recognized as a model for biological research in the field of regeneration, stem cell biology, study of their proliferation and differentiation, as well as the regulatory mechanisms of morphogenetic processes. The genome of the planarian *Schmidtea mediterranea* was fully sequenced which provides an excellent opportunity for the analysis of this species at the molecular biological level. It has become clear that many of the genes controlling regeneration in planarians are conserved in Metazoa [4–6], which further increases the relevance of studying the processes of posttraumatic recovery of the planarians under the impact of radiation.

The effects of low-intensity electromagnetic fields (EMF) and low and medium doses of γ-radiation on these processes are not well studied, and little is known about the response of stem cells to such exposures.

We have previously conducted a comparative analysis of preparation methods for *S. mediterranea* samples for flow-cytometric analysis of proliferative activity [7]. The aim of this study is to analyse the short- and long-term effects of radiation on the regeneration and proliferation of directly exposed *planarian*. We compared the proliferation activity of irradiated and control planarians after γ-irradiation at dose 10 Gy and low-intensity electromagnetic radiation (EMR) with mobile phone parameters. Proliferative activity was evaluated by flow cytometry with widely applied method that allows determining the proportion of cells at different phases of the cell cycle, including S- and G2/M-phases, in which there may be cycling, but not resting cells. Regeneration activity was quantified as the ratio of area of blastema to the entire body of planarian using computerised morphometry.

## 2. Material and methods

### *2.1. Design of irradiation*
The planarians of 10 mm were selected, decapitated and divided into three groups. The first group of animals was irradiated for 180 min at the laboratory facility, continuously generating EMR with a frequency of 900 MHz and an energy flux density of 100 $\mu W/cm^2$. The second group was irradiated at the therapeutic installation 'Luch-1' (Latvia, $^{60}Co$) at a dose of 10 Gy (dose rate 94 Gy/min). The selected dose is less than 10% of $LD_{50}$ for this species (about 60–80 Gy).The control was in the same conditions, but without irradiation. The Index of regeneration of exposed and control planarian was evaluated at 4$^{th}$ day after decapitation and exposure. The analysis proliferation activity was performed at 6$^{th}$, 8$^{th}$ and 24$^{th}$ hours.

### *2.2. Cell Cycle Analysis*
For cell cycle analysis we used a well-known technique by staining cells with propidium iodide (PI) in the presence of RNase [8]. However, before this, we applied chemical disaggregation of cells in the presence of citric acid (0.1 M) for 10 min at room temperature. Then cells were slowly fixed in chilled ethanol, centrifuged and incubated with staining solution in the dark for 30 min at room temperature, filtered through 40-μm nylon filter. Finally, we used FACS Calibur flow cytometer (BDIS, USA) and CellQuestPro [9] to determine light scattering and intensity of PI fluorescence (585±42 nm wavelength).

Using ModFit 3.1 program (BDIS, USA) a region of cells was identified according to the intensity of forward and side light scattering, then the debris and conglomerates of cells (including so-called doublets) were excluded from the analysis considering the parameters FL2-H/FL2-W that characterized the maximum intensity and duration of the fluorescence signal of each event. Histograms of the cell distribution by the integrated fluorescence intensity of PI were built after gating to determine the proportion of cells at different phases of cell cycle. Proliferation activity was estimated as the total fraction of cells at the S- and G2/M phases of the cell cycle.

### *2.3. The analysis of regeneration*
The analysis of regeneration was performed in a small sample of the planarian *S. mediterranea* (n=48 for γ-irradiation and n=57 for EMR) by computer morphometry in vivo. The Index of regeneration of exposed and control planarian was evaluated at 4$^{th}$ day after decapitation and exposure [10]. The blastemal growth-rate in controls and irradiated flatworms was established after decapitation by scoring the old pigmented and newly grown non-pigmented cells and quantified as the ratio of area of blastema to the entire body of planarian.

### *2.4. Statistical method of analysis*
Three independent series of experiments were carried out for analyze of proliferation activity. In the experimental groups, 50 samples were analyzed (5 planarians per sample), in the control group – 36 samples. The number of cells in each sample was at least 200 thousand. The calculation was made in

the Goryaev's chamber. Statistical processing of the results was performed according to the Kruskal-Wallis test.

Statistical processing of the results of regeneration activity was performed according to the Mann-Whitney U-test.

## 3. Results and discussion

### *3.1. The analyses of the proliferative activity in S. mediterranea*

The dynamics of changes in the proliferative activity of cells of regenerating planarians through 6, 8 and 24 h after γ-irradiation at a dose of 10 Gy compared to that after exposure to EMR (900 MHz , 100 mW/cm$^2$, 180 min) are presented in table 1.

It is known that mitotic activity has the first maximum in 6 h after decapitation and affects the whole body of the animal, the second maximum is observed locally at the site of damage 48–72 hours after it [11]. The first mitotic wave is determined mainly by G2 neoblasts, the second – neoblasts at G0 and G1 phases of the cell cycle at the time of damage. According to our data, the proliferative activity of planarian cells in the first six and eight hours after EMR exposure did not differ from that in control (nonradiated regenerating) animals. 24 h after EMR exposure, the number of neoblasts at the S and G2/M phases decreased by 1.5 times compared to the control. A greater drop in proliferative activity was observed in γ-irradiated planarians compared to the effect of EMR. Thus, 24 h after γ-irradiation, the fraction of S-phase cells decreased by 1.8 times, and in G2/M-phases – by 5.3 times compared to the control. At this period, the total number of (S + G2/M) cells characterizing proliferative activity after ionizing radiation exposure decreased by 2.8 times compared to the control and was lower by 1.8 times than after the action of EMR. It should also be noted that a decrease in proliferation was observed as early as 8 hours after γ-irradiation as opposed to EMR exposure.

According to the results of two-way ANOVA, there are highly significant effects of γ-irradiation ($p=1.54\cdot10^{-8}$) and time following exposure ($p=0.0009$) on the proliferation in Planarian. The contribution of time after irradiation and the interaction of these factors was significant ($p=2.93\cdot10^{-5}$). The effects of EMR ($p=0.02$) and time following exposure ($p=0.0001$) on the proliferation in Planarian were significant too.

Figure 1 shows the percentage of cells in the controls and irradiated samples of *S. mediterranea* planarian.

It can be seen that 24 hours after exposure at γ-rays and EMR the index of proliferation established as the total amount of cells at S and G2/M phases, was 2.8 and 1.8 lower than that in controls respectively.

Our data therefore provide the first experimental evidence for the effects of medium-dose γ-irradiation and low-intensity EMF on regeneration in planarians. It proves, in particular, the need for careful monitoring of anthropogenic load levels near the sources of both ionizing and non-ionizing (radio-frequency) radiation.

### *3.2. The analyses of the regeneration activity in S. mediterranea*

Figure 2 shows the index of regeneration activity in planarians on the fourth day after decapitation and exposure at low-intensity EMF and γ-rays at dose 10 Gy.

According to our results, the regeneration activity is significantly compromised in the γ-irradiated groups. It should also be noted that the results of our study were obtained within the dose far below the semi-lethal doses for *Planaria* of 60-80 Gy. The regeneration activity after EMR exposure did not significantly differ from that in controls.

IOP Conf. Series: Materials Science and Engineering **487** (2019) 012018 doi:10.1088/1757-899X/487/1/012018

**Table 1.** The distribution of *S. mediterranea* cells across different phases of cell cycle at 6[th], 8[th] and 24[th] hours after γ-radiation and EMR exposures.

| | | | | | |
|---|---|---|---|---|---|
| | | **γ--irradiation** | | | |
| Dose (Gy) | $N^a$ | Proportion of cells at different phases (%) ± SE | | | |
| | | *G1/G0* | *S* | *G2/M* | *S+G2/M* |
| | | 6 h after irradiation | | | |
| 0 | 9 | 74.3±1.0 | 18.6±1.3 | 8.0±0.6 | 26.5±1.0 |
| 10 | 9 | 75.0±1.1 | 17.2±1.0 | 7.8±0.4 | 25.0±1.1 |
| Kruskal-Wallis test[b], df=1 | | 0.33 | 1.04 | 0.16 | 1.65 |
| $P^c$ | | 0.57 | 0.31 | 0.69 | 0.20 |
| | | 8 h after irradiation | | | |
| 0 | 8 | 68.3±2.7 | 17.7±1.6 | 14.0±2.6 | 31.8±2.7 |
| 10 | 7 | 77.2±1.6 | 4.2±2.4 | 18.6±1.7 | 22.8±1.6 |
| Kruskal-Wallis test, df=1 | | 3.87 | 7.73 | 1.93 | 3.87 |
| *P* | | 0.05 | **0.01** | 0.16 | 0.05 |
| | | 24 h after irradiation | | | |
| 0 | 19 | 70.3±1.4 | 14.2±0.8 | 15.9±1.2 | 30.0±1.5 |
| 10 | 6 | 89.1±0.7 | 7.8±0.8 | 3.0±0.9 | 10.9±0.7 |
| Kruskal-Wallis test, df=1 | | 13.15 | 10.95 | 13.15 | 13.15 |
| *P* | | **0.0003** | **0.0009** | **0.0003** | **0.0003** |
| | | Two-factor analysis of variance (two-way ANOVA) | | | |
| Contribution of gamma irradiation to the effect | | $2.58 \cdot 10^{-8}$ | $5.38 \cdot 10^{-8}$ | 0.035 | $\mathbf{1.54 \cdot 10^{-8}}$ |
| Contribution of time after irradiation | | 0.0007 | $1.43 \cdot 10^{-6}$ | $6.28 \cdot 10^{-6}$ | **0.0009** |
| Interaction of factors | | $1.46 \cdot 10^{-5}$ | 0.0003 | $3.06 \cdot 10^{-6}$ | $\mathbf{2.93 \cdot 10^{-5}}$ |
| | | **Electromagnetic exposure** | | | |
| Dose (Gy) / Exposure time (min) | $N^*$ | Proportion of cells at different phases (%) ± SE | | | |
| | | *G1/G0* | *S* | *G2/M* | *S+G2/M* |
| | | 6 h after irradiation | | | |
| 0 (control) | 9 | 74.3±1.0 | 18.6±1.3 | 8.0±0.6 | 26.5±1.0 |
| 180 min | 9 | 75.4±1.4 | 15.5±1.1 | 9.1±0.5 | 24.7±1.4 |
| Kruskal-Wallis test[b], df=1 | | 1.22 | 3.30 | 2.14 | 2.69 |
| $P^c$ | | 0.27 | 0.07 | 0.14 | 0.10 |
| | | 8 h after irradiation | | | |
| 0 (control) | 8 | 68.3±2.7 | 17.7±1.6 | 14.0±2.6 | 31.8±2.7 |
| 180 min | 10 | 66.5±1.6 | 17.7±0.7 | 15.8±1.1 | 33.5±1.6 |
| Kruskal-Wallis test, df=1 | | 0.07 | 0.20 | 0.96 | 0.07 |
| *P* | | 0.79 | 0.66 | 0.33 | 0.79 |
| | | 24 h after irradiation | | | |
| 0 (control) | 19 | 70.3±1.4 | 14.2±0.8 | 15.9±1.2 | 30.0±1.5 |
| 180 min | 10 | 80.0±1.6 | 9.7±1.4 | 10.3±1.5 | 20.0±1.7 |
| Kruskal-Wallistest, df=1 | | 9.73 | 5.48 | 5.48 | 9.74 |
| *P* | | **0.002** | **0.019** | **0.019** | **0.002** |
| | | Two-factor analysis of variance (two-way ANOVA) | | | |
| Contribution of electromagnetic exposure | | 0.04 | 0.01 | 0.47 | **0.02** |
| Contribution of time after irradiation | | $3.68 \cdot 10^{-5}$ | $1.62 \cdot 10^{-6}$ | 0.0004 | **0.0001** |
| Interaction of factors | | 0.003 | 0.155 | 0.017 | **0.003** |

[a] sample size

[b] Kruskal-Wallis test values (df=1) for comparison with the control group

[c] the probability of differences adjusted for multiple comparison Bonferroni

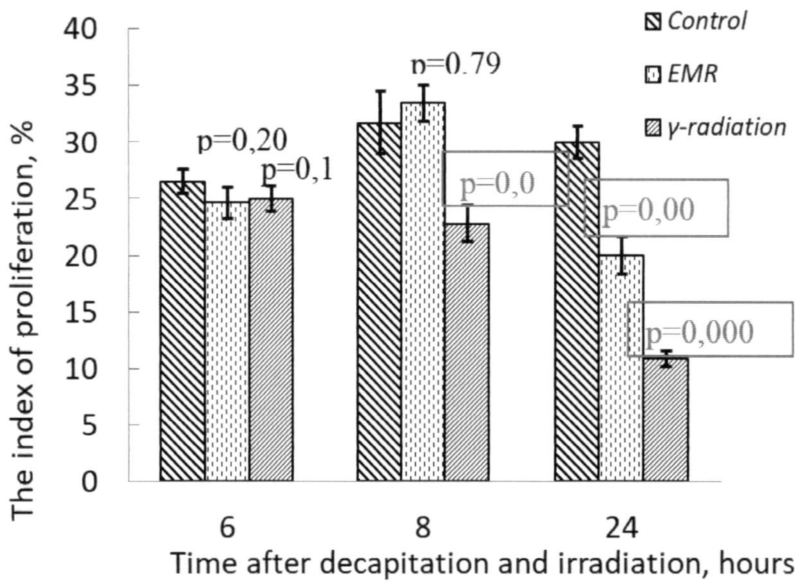


**Figure 1.** The percentage of cells at the S- and G2/M-phases of cell cycle in the samples of *S. mediterranea* regenerating planarians in control and exposed groups.

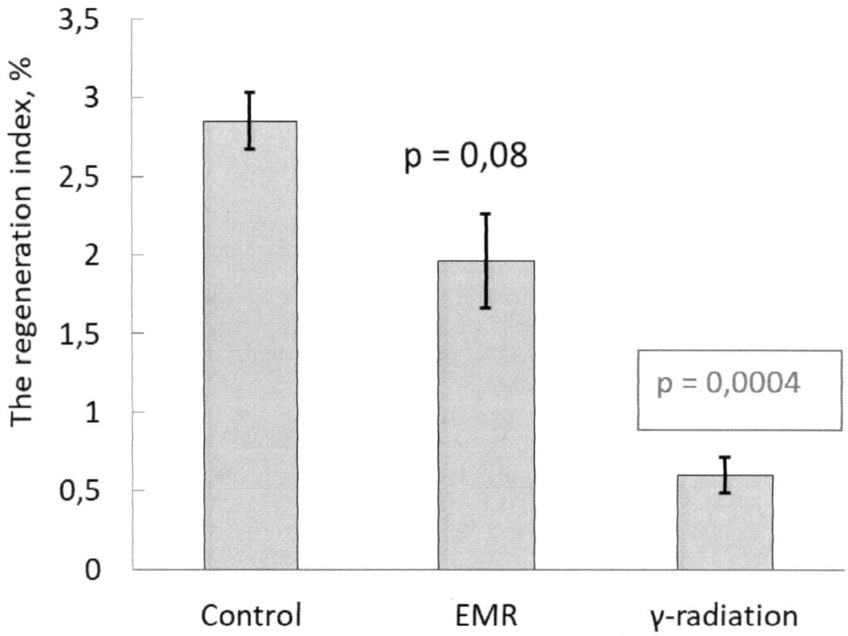


**Figure 2.** The index of regeneration in the controls and samples of *S. mediterranea* planarians exposed to EMR (900 MHz, 100 μW/cm$^2$, 180 min) and γ-radiation at dose of 10 Gy.

## 4. Conclusions

The decrease in the proliferative activity of the *S. mediterranea* planarian cells after EMR exposure was found to be less pronounced than that after γ-irradiation and not be reflected in the regeneration. This indicates the leveling of the EMR effect in the more distant periods, possibly due to the subsequent stimulation of the neoblast proliferation, the increase in migration activity or the intensification of the differentiation processes.

Using of planarians as the object of study allowed us to analyze the effects of ionizing and non-ionizing radiation on both cellular and organismic levels.

Our data therefore provide the first experimental evidence for the effects of medium-dose γ-irradiation on regeneration in planaria which is attributed to radiation-induced compromised proliferation activity.

**Acknowledgments**
Author acknowledges support from the MEPhI Academic Excellence Project (No. 02.a03.21.0005). This work was partially supported by the Russian Foundation for Basic Research (18-48-400010) to E.I.S.

## References

[1] Baguna J 2012 The planarian neoblast: the rambling history of its origin and some current black boxes *Int. J. Dev. Biol.* **56** 19–37
[2] Krishna S, Palakodeti D and Solana J 2018 Post-transcriptional regulation in planarian stem cells *Semin Cell Dev Biol.* pii: S 1084-9521(17)30204-5
[3] Rink J C 2013 Stem cell systems and regeneration in planarian. *Dev Genes Evol.* **223** 67–84
[4] Fumagalli M R, Zapperi S and La Porta C A M 2018 Regeneration in distantly related species: common strategies and pathways. *NPJ Syst Biol Appl.* **4** 5
[5] Kenny N J, de Goeij J M, de Bakker D M, Whalen C G, Berezikov E and Riesgo A 2018 Towards the identification of ancestrally shared regenerative mechanisms across the Metazoa: A Transcriptomic case study in the Demosponge Halisarca caerulea *Mar Genomics* **37** 135–47
[6] Wenemoser D, Lapan S W, Wilkinson A W, Bell G W and Reddien P W 2012 A molecular wound response program associated with regeneration initiation in planarians *Genes Dev* **26** 988–1002
[7] Markina E S, Matchuk O N and Sarapultseva E I 2017 Comparative analysis of methods of sample preparation of *Schmidtea mediterranea* flatworms for cytometric analysis *Technogenic systems and environmental risk. Book of abstracts of the first int. scientific conf.* ed A A Udalova (Obninsk: NRNU MEPhI) p 118
[8] Ermakov A M, Ermakova O N, Kudravtsev A A and Kreshchenko N D 2012 Study of planarian stem cell proliferation by means of flow cytometry *Mol Biol Rep.* **39** 3073–80
[9] Matchuk O N, Zamulaeva I A, Kovalev O A and Saenko A S 2013 Radioresistance mechanisms of side population cells in mouse melanoma cell line B16. *Tsitologiia* **55**(8):553-9
[10] Tiras K P and Aslanidi K B 2016 Two populations of pluripotent stem cells in planarians Girardia tigrina *Biochemistry (Moscow) Supplement. Series A: Membrane and Cell Biology* **10** 46–52
[11] Wenemoser D and Reddien P W 2010 Planarian regeneration involves distinct stem cell responses to wounds and tissue absence *Dev Biol.* **344** 979–91

# Mixture toxicity of different metal ions in cytogenetic biotests

**B I Synzynys[1], L N Ulyanenko[2], N V Amosova[1], S V Pyatkova[1] and O A Момот[1]**

[1] Obninsk Institute for Nuclear Power Engineering, National Research Nuclear University MEPhI, Obninsk, Russia
[2] A. Tsyb Medical Radiological Research Centre, National Medical Research Radiological Centre, Ministry of Health of the Russian Federation, Obninsk, Russia

E-mail: synzynys@obninsk.ru

**Abstract**. The separate and combined impacts of aluminium and selenium ions on different barley cultivars and onion (*Allium cepa* L.) seedling as well as ions copper, zinc and nickel on onion have been studied. The experiments have been performed in several series with different solutions of ions: aluminium have been used in concentration of 0.5 mg/l (Russian MPC for cultural-domestic water), ions of Fe, Mn, Se, Ag and their combinations with ions of Al have been used in concentrations corresponding to the values of Rus MPC for cultural-domestic water. At the same condition were chosen the concentrations of copper, zinc and nickel. The phytotoxic and genotoxic effects (as mitotic index – MI and aberrant cell frequency – ACF) were estimated. It was showed that aluminium ions promote higher phytotoxic and genotoxic effects, but in combined with another ions the negative effects on barley and *Allium cepa* reduced. This confirms that the universal mechanism the interaction between two metals (aluminium and other) and their detoxification effect in plant. The essential metals Cu and Zn and the toxic metal nickel provide a strong cytogenetic effect which results to ACF increasing and MI suppression. We assume that the mechanisms of detoxication should be initiated by metals themselves.

## 1. Introduction

Metal ions in various combinations are always present in natural water bodies used for fishing, cultural and domestic purposes. As a rule, regulatory actions of each ion concentration do not consider their combined impact on people, aquatic animals and plants and it probably appears as synergism, antagonism and additive effects [1]. It demands the study of these reactions in biotests and regulation of water pollution.

Aim – detection and qualitative cytogenetic analysis of a combined impact of metal ions.

## 2. Materials and methods

Cytogenetic methods of biotesting are used to assess the direct and combined effects of metal ions in water. The mitotic index (MI) and the aberrant cell frequency (ACF) are determined in the root meristem of barley and onion (Allium-test) seedlings during their sprouting in metal solutions [2]. Aluminum which at the concentration of 0.5 mg/l (MPC for cultural-domestic water) and pH = 4.5 has a strong cytogenetic effect is taken as the primary acting agent [3]. Coupled with Al, ions of Fe, Mn, Se and Ag and their combinations have been used in concentrations corresponding to the Russian MPC values for cultural and domestic water.

Other experiments with Stuttgarten Risen onion seedlings made it possible to assess the cytogenetic changes after sprouting in zinc, copper and nickel solutions in concentrations (according to the metal ion) which correspond to MPC values for cultural-domestic water. The coupling factor (antagonism or synergism) is found from $K=S/S_1+S_2$, where S is the measured ACF value (here minus from a reference value); $S_1$ and $S_2$ are the corresponding ACF values under the impact of each metal particularly (minus reference values). The statistical data processing from three independent experiments is realized using the Stat Graf program.

## 3. Results

Table 1 presents the experimental results for seedlings of two barley cultivars and Stuttragten onion; the data are assessed from the mitotic index (MI) of root meristem cells and in Allium-test.

**Table 1.** Mitotic index in the root meristem of barley and onion seedlings.

| Species | MI (%) | | | |
|---|---|---|---|---|
| | **Blank** | **$Al^{3+}$** | **$Se^{2+}$** | **Al+Se** |
| Bios, barley | 6.5±0.3 | 2.5±0.2 | 4.3±0.2 | 5.5±0.3 |
| | **Blank** | **$Al^{3+}$** | **$Fe^{2+}$** | **Al+Fe** |
| Bios, barley | 8.4±0.6 | 3.2±0.3 | 8.9±0.7 | 5.4±1.0 |
| Getman, barley | 10.0±0.3 | 6.1±0.2 | 9.3±0.8 | 7.2±0.3 |
| | **Blank** | **$Al^{3+}$** | **$Mn^{2+}$** | **Al+Mn** |
| *Allium cepa* | 8.2±0.6 | 5.1±0.5 | 6.2±0.7 | 7.4±0.7 |
| | **Blank** | **$Al^{3+}$** | **$Ag^{2+}$** | **Al+Ag** |
| *Allium cepa* | 10.1±0.9 | 5.9±0.5 | 6.6±0.9 | 7.0±0.6 |

It was earlier established with wheat seedlings that $Al^{3+}$ ions found in the water environment at pH = 4.5 have a profound cytogenetic effect [3]. This regularity is also confirmed by barley and onion seedlings: in the concentration used the ions of Al suppress the mitotic index of Bios barley which is sensitive to its effects by 62% and the mitotic index for cells of the resistant Getman variety goes downs by 39 % only. For the root meristem cells of onion the degree of suppression in various experiments ranges from 38 to 42%. Ions of other metals (Se, Mn, Ag) also suppress the mitotic activity of meristem cells, however, in case of a combined (with aluminum) effect the mitotic activity rises as compared to the action of Al alone. The same effect is observed in a combined action of aluminum and ferrous iron, the latter having no cytotoxic effect. One of the expected reasons for suppressing the plant cell division is the disrupted chromosome integrity and mitoses caused by the indirect impact of various oxide radicals [4]. The aberration rate has been studied to test this assumption of disruption in the genetic apparatus of dividing cells of the root meristem seedlings. These results are presented in table 2.

It is shown that selenium characterized by pronounced antioxidant properties has demonstrated its protective effect for Bios barley in combination with aluminium. In this case the antagonism factor is 0.66. If aluminium and ferrous iron ions are present in the solution simultaneously, there is a pronounced effect of antagonism in case of Bios (K = 0.47) which is more sensitive to Al ions as compared to the resistant variety Getman (K = 0.77). Even more pronounced antagonism (K = 0.20) is observed in a combined impact of bivalent manganese and silver ions on onion seedlings.

The following series of experiments allowed the separate and combined effects of essential (Cu, Zn) and certainly toxic (Ni) metals (table 3) to be studied. The problem consisted in observing the highest effect of antagonism from the cytogenetic indices under investigation.

**Table 2.** Cytogenetic effect of metal ions and their combinations on barley and onion seedlings.

| Species | ACF (%) | | | | K |
|---|---|---|---|---|---|
| | **Blank** | **$Al^{3+}$** | **$Se^{2+}$** | **Al+Se** | |
| Bios, barley | 1.6±0.4 | 6.3±0.6 | 3.1±0.4 | 4.1±0.6 | 0.66 |
| | **Blank** | **$Al^{3+}$** | **$Fe^{2+}$** | **Al+Fe** | |
| Bios, barley | 1.7±0.3 | 6.8±0.4 | 1.5±0.4 | 4.1±0.7 | 0.47 |
| Getman, barley | 0.7±0.4 | 3.4±0.3 | 0.7±0.5 | 2.8±0.4 | 0.77 |
| | **Blank** | **$Al^{3+}$** | **$Mn^{2+}$** | **Al+Mn** | |
| *Allium cepa* | 1.4±0.8 | 4.1±1.4 | 1.3±0.6 | 1.9±0.9 | 0.20 |
| | **Blank** | **$Al^{3+}$** | **$Ag^{2+}$** | **Al+Ag** | |
| *Allium cepa* | 1.2±0.8 | 4.0±1.9 | 3.5±1.2 | 2.2±1.1 | 0.20 |

ACF – aberrant cell frequency, K – antagonism factor.

**Table 3.** Genotoxic effect of Cu, Zn and Ni ions on *Allium cepa* root meristem [2].

| Version | Number of ana-telophase cells | Aberrant cell frequency (%) | Mitotic index (%) | Type of chromosome aberration |
|---|---|---|---|---|
| Cu | 314±4 | 6.56±0.82 | 7.24±0.46 | m, f |
| Zn | 333±3 | 5.16±0.90 | 8.77±0.41 | m, f, 3p |
| Ni | 438±4 | 7.76±1.37 | 9.96±0.40 | m, f, g |
| Cu + Zn | 391±4 | 3.03±0.20 | 8.97±0.45 | m, f |
| Zn+ Ni | 324±3 | 3.04±0.39 | 7.35±0.34 | m, f |
| Ni + Cu | 177±2 | 3.83±0.80 | 6.32±0.42 | m, f |
| Cu +Zn+ Ni | 347±4 | 3.85±0.54 | 7.13±0.40 | m, f |
| Reference | 781±8 | 0.58±0.23 | 15.12±0.73 | M |

Cu – 1 mg/l, Zn – 1 mg/l, Ni – 0.02 mg/l.

As seen from table 3, both the essential metals (copper and zinc) and the plant toxic nickel used in given concentrations (MPC for cultural-domestic water bodies according to the Russian health legislation) cause a strong cytogenetic effect which results in the certain ACF increase and MI suppression. A significant decrease of cytogenetic as well as cytotoxic effects can be achieved by combining these metals in germination solutions. However, the damaging effect of all three metals could not be fully eliminated even in their combined action. There were no tripolar mitoses (as most destructive to cell activity) under the action of essential metals as well as their combinations with admittedly toxic nickel.

## 4. Discussion

In 1909 the German agrochemist E. Mitcherlich had published the paper formulating ‘the law of additive effects of environmental factors’. The law says: the well-being of species, population, organisms depends not only on a particular, though a limiting factor, but simultaneously on the totality of ecological factors. The discovery of any law of nature, including this one, can be considered as a certain key to one of the intentions of the Creator-the Author of the Universe. In this instance it can be assumed that in planning the Universe creation in its current form the Creator knew in advance that it would be based on 100 chemical elements. Supposing that the initial optimum amount of each this element is exceeded by 1% (in their biological effect) it would result in almost complete disappearance of life, plant life in particular. To prevent the expected adverse developments, the existence of living things is based on special defense mechanisms. Being under a pleioptropic genetic control and, hence, considering a vast variety of their manifestations, these mechanisms should cancel and reduce the effect of increased values (higher than the optimal ones) of aluminum concentrations. The mechanisms of detoxication should be initiated by metals themselves. If one of the metals suppresses the plant growth and development, the neighbor metal would cancel or reduce the adverse effects. The preliminary results on a combined action of various metals or fluorine on barley and wheat were presented in 2009 in Syktyvkar at the 100 anniversary of E. Mitcherlich law [5]. Antagonistic manifestations of induction effects of different chromosome aberrations followed by an increased mitotic cell activity were demonstrated at the cytogenetic level. Recently similar regularities have also been found by Belgian scientists in the experiments with a simultaneous impact of four metals (copper, nickel, cadmium and zinc) on wheat seedlings [6].

**References**

[1] Synzynys B I, Momot O A, Ulyanenko L N, Amosova N V, Pyatkova S V and Lavrentieva G V 2017 Mechanisms of biological action of metal ions *Proc. Int. Conf. 'Technogenic systems and ecological risk'* (Obninsk: INPE NRNU MEPhY) pp 164–5

[2] Ulyanenko L N, Reva E V and Synzynys B I 2017 Cytogenetic effects in Allium cepa L. resulted from separated and combined exposure to Cu, Zn and Ni *Agricultured Biology* **52** 182–90

[3] Bulanova N V, Synzynys B I and Koz'min G V 2001 Aluminum induces chromosomal aberrations in root meristem cells of wheat *Russian. J. Genet.* **37** 1455–8

[4] Skugoreva S G, Ashihmina T Ya, Fokina A I and Lyalina E I 2016 Chemical grounds of toxic effect of heavy metals (a review) *Theoretical and Applied Ecology* **1** 4–13

[5] Synzynys B I, Amosova N V, Reva E I, Polyakova I V, Simakova I M and Netsvet N V 2009 100 anniversary of E.A. Mitcherlich law for a combined action of ecological factors: principles of a combined action of essential and toxic metals and radiation on plants on the cytogenetic level *Biologicheskie effekty malykh doz ioniziruyushchey radiatsii i radioaktivnoe zagryaznenie okruzhayushchey sredy* (Syktyvkar) pp 371–3

[6] Versieren L, Evers S, Smolders E, De Schamphelaere K and Blust R 2016 Mixture toxicity and interactions of copper, nickel, cadmium and zink to barley at low effect levels: something from nothing? *Environmental toxicology and chemistry* **35** 2483–92

IOP Conf. Series: Materials Science and Engineering **487** (2019) 012020 doi:10.1088/1757-899X/487/1/012020

# Peptide regulation of plant and animal morphogenesis: general mechanisms and specificity of action

**Kh Tiras[1,2], U Vorobeva[1,2], K Novikov[3], V Voeikov[3], V. Emelyanenko[2,4], G Davidova[2], A Skripnikov[3,5]**

[1]Puschino State Institute of Natural Sciences, Puschino, Russia
[2]Institute for Theoretical and Experimental Biophysics of RAS, Moscow, Russia
[3]Lomonosov Moscow State University, Biological faculty, Moscow, Russia
[4]Institute for Biological Instrumentation of RAS, Moscow, Russia
[5]M Shemyakin - Yu Ovchinnikov Institute of Bioorganic Chemistry of RAS, Moscow, Russia

E-mail: tiras1950@yandex.ru

**Abstract.** Regulatory peptides are one of the main signaling molecules that provide intercellular interactions in biological systems. Thus, regulatory peptides are known as morphogens - biologically active substances involved in the regulation of the morphogenesis of animals and plants. Until now, the interaction of plant and animal peptides and, in particular, the 'cross' effect of plant peptides on the morphogenesis of animals and vice versa has not been sufficiently investigated.

The present work is devoted to the study of the effect of growth regulator peptide (CLV 3) of *Arabidopsis* on the morphogenesis of animals - freshwater flatworms – planaria. This peptide is an effective regulator of growth and differentiation of plant roots in low and ultra-low concentrations.

Various biophysical methods have shown that this plant peptide at a concentration of $10^{-9}$M is also an effective morphogen of animals. At the same time, the specificity of the action of peptides on 'unrelated' biological targets has been found. The peptide CLV 3 stimulates on the morphogenesis of plant roots in concentrations up to $10^{-12}$M, while for planarian regeneration it acts only at concentrations up to $10^{-9}$M, at which it does not show a noticeable effect on root plant growth. When this peptide was applied to the of stem human stem cells Th culture, there was no effect observed.

The results of the study confirm the thesis that there is a general biological level and peptide regulation mechanisms that are similar for animals and plants, and also reveal the nature of the specifics of the 'direct' and 'cross' action of this mechanism.

## 1. Introduction

The purpose of this work is to study the effect of the biologically active plant dodecapeptide CLV3 on the process of regeneration of flatworms – planaria as an example of the possible influence of peptide systems of plant regulation on the processes of animal morphogenesis.

IOP Conf. Series: Materials Science and Engineering **487** (2019) 012020 doi:10.1088/1757-899X/487/1/012020

Morphogenetically active peptides can be an important tool for molecular interaction between representatives of different biological taxa living in close proximity. An example of direct contact of plants and animals can serve as flat worms and plants in the roots of which they inhabit.

Planarian regeneration is regulated by various biologically active peptides of animals, being morphogenic, therefore flatworms were used as a regenerative model for the study of morphogenesis. It has been shown that different peptides are able to control the processes of proliferation and regeneration in extremely low concentrations (up to $10^{-15}$M), which allows us to consider peptides-morphogenes as one of the finest instruments of molecular control [1–4].

The peptide pCLV3 is expressed in the stem portion of *Arabidopsis thaliana*, influencing the growth and development of the terrestrial part and plant roots in concentrations up to $10^{-9}$M – $10^{-12}$M [5].

In this paper, the effect of this peptide on morphogenesis was studied during regeneration of the cephalic end of the body of planar worms during five days of regeneration using non-invasive biophysical recording methods. During the first 24 hr after the cutting, ultra-weak photon emission (SWFE) was recorded in the planarians, and after 72 hr and after 120 hr of regeneration the epimorphic growth of the head end (blastema) and the rest of the planarian body were recorded by the method of IDM [6, 7].

## 2. Materials and methods

The work was carried out on an asexual laboratory race of the planaria *Girardia tigrina*, which were kept in room conditions at a temperature of 20 – 21 °C. The animals were fed weekly with *Diptera* larvae. Before the experiments, the planarians had starved for a week.

The dodecapeptide CLV3, having the following sequence of amino acid residues RTVPSGPDPLHH, was provided by the Laboratory of Peptide Chemistry of the Institute of Bioorganic Chemistry RAS. MM Shemyakin and YuA Ovchinnikov. The dodecapeptide CLV3 was prepared by N-α-FMOC solid phase synthesis and purified by HPLC using a standard procedure. The identity and purity of the synthesized peptide at the level of 95% is confirmed by reversed-phase HPLC, and also by means of MALDI-TOF/TOF-mass spectrometry.

### *2.1. Superweak photon emission (SWFE)*

The registration of SFE was carried out with the help of the Biotox 7a luminometer. The sample is equipped with a 9750QB/1 photomultiplier (9750QB/1, EMI Electronics LTD, Middlesex, UK) with a spectral sensitivity region of 380-710 nm, with a maximum in the blue-green region, 420-500 nm. The measurements were carried out at a temperature of 21.0 ± 0.2 ° C, in the photocount counting mode with an interval of 1c. Further analysis of the results was carried out according to the previously described method [7].

A group of 30 planarians immediately after decapitation was placed in a scintillation counting bottle in 10 ml of an aqueous solution of lucigenin ($10^{-9}$M), and the registration of SSE was started. In special experiments, the registration of SWFE in planarians in the control and experimental groups was carried out in parallel on two Biotox-7a luminometers.

### *2.2. Intravital digital morphometry (IDM)*

The method is based on the registration of a photocontrast between the old (pigmented) and new (uncoated pigment) parts of the body of the regenerating planaria. To obtain the planar images, a Stemi 2000C binocular computer microscope equipped with an AxioCam MRc (Zeiss) video camera was used. Regeneration was caused by amputation of the head part of the planar body and the regeneration was recorded at 72 and 120 hrs after transection.

The digitized images of the regenerating planarians were processed in the Plana 5.0 program, which allows analysis of electronic images of the planaria and determination of the areas of the projection of the regenerant body and blastema. As a criterion for regeneration, the ratio of the area of the blastema to the total area of planaria was applied on the certain hours of regeneration [6].

*2.3. Definition of the mitotic index (MI)*
The mitotic index during regeneration was determined by counting the mitotic figures in the suspension of planar cells [8]. The number of metaphases per 1000 cells (mitotic index) was counted within 24 hours after transection at colchicine exposure. To visualize the mitotic figures, Hoehst 33342 dye was used. Metaphases were recorded with the AxioVert 200M (Zeiss) fluorescent microscope.

*2.4. Cytotoxicity assay*
Cytotoxicity was analyzed using the postnatal dental pulp stem cells *Th* (DPSC) [9]. For the experiment, DPSC were plated in wells of a 96 well plate at a concentration of 25,000 cells/cm$^2$ in DMEM/F12 medium containing 5% FBS serum. After 18 hours, the medium was removed and $10^{-7}$-$10^{-11}$M peptide solution in DMEM/F12 medium containing no serum attachment factors was added. As control of the passage of the MTT-test, 10% dimethyl sulfoxide (DMSO) was added to the cell culture medium. As a general control, cells cultured in DMEM / F12 medium containing no serum attachment factors were used. In the second case, the medium with the peptide was changed every 24 hr.

To evaluate the proliferation, death of human stem cells and the of the material, a MTT test was used based on the direct correlation of the number of living cells and the intensity of the metabolism of MTT (3-(4,5-dimethylthiazolyl-2)-2,5-diphenyltetrazolium bromide) insoluble dark-colored formazan soluble in dimethylsulfoxide under the action of dehydrogenases [10].

*2.5. Statistical analysis*
The statistical analysis of the results of the mitotic index determination was carried out using the Mann-Whitney criterion for small groups. The results of the IGM image analysis were carried out using the SygmaPlot 9.0 software package. The mean value, the mean error, the confidence interval for the significance level of 0.95 was determined. The comparison of the averages was carried out using the Student's t-test.

## 3. Results and discussion
In this work, two non-invasive biophysical approaches were used to register the regeneration processes. In the first 24 hr after the transection, we recorded own planarian superweak photon emission (SWFE), and then on the same group of animals – epigenetic regeneration was recorded 72 and 120 hr after decapitation by IDM. In our study we used the registration of SWFE as a remote test for the lifetime detection of ROS as a regulator of cell proliferation and their differentiation, as well as an indicator of metabolic activity during regeneration, within the first 24 hr after transection.

During regeneration, in many animals, including planarians, the level of generation of ROS increases [11]. This process of sequential four-fold one-electron reduction of oxygen is accompanied by generation of electron excitation energy and leads to emission of photons of super-weak intensity which is detected by highly sensitive photomultipliers. In our study we used the registration of SWFE as a remote test for the lifelong detection of ROS.

After operations, all fragments of control and experimental animals were placed in a luminometer for 24 hr. Further, the blastema of the decapitated planaria of the experimental and control groups was recorded at 72 and 120 hr after the operation by the IDM method.

In a series of the very first test experiments on the study of the biological activity of CLE-peptides, dodecapeptides were synthesized, based on the genomic data on the sequence of amino acids in the region of the conserved domain of 26 representatives of the CLE *A. thaliana* family [12].

In this study, we used one of the most studied and characterized CLE peptides of *A. thaliana* the dodecapeptide CLV3. It was previously established that when the peptide CLV3 is added to the culture medium at concentrations of $10^{-6}$M to $10^{-8}$M, the growth of the main root of the Arabidopsis seedlings is significantly suppressed, and at the same time the formation of lateral roots is stimulated. At a concentration of $10^{-12}$ M peptide, CLV3 causes stimulation of the growth of the main root. At this concentration dodecapeptide CLV3 reduces the number of lateral roots [5].

IOP Conf. Series: Materials Science and Engineering **487** (2019) 012020 doi:10.1088/1757-899X/487/1/012020

The study of SWFE was carried out in parallel in two groups (experimental and control) of 30 decapitated animals on two Biotox-7a luminometers. The results are shown in Figure 1. Immediately after the cutting, an intensive burst of SWFE is observed. Then, within one to two hours after the transection, the relaxation of SWFE is registered to the level of the intact planaria [7].

The CLV3 peptide in the first two hours after the decapitation operation accelerated the relaxation of the level of SWFE twice as compared with the control group. The main peak of SWFE was observed 12 hours after the decapitation operation: the level of SWFE was twice as high under the action of the peptide, which was confirmed in 5 series of experiments. This peak coincides in time with the peak of proliferation of stem cells of planarians – neoblast [13]. This assumption was confirmed in experiments on the study of the mitotic index of planaria during the first day of regeneration.

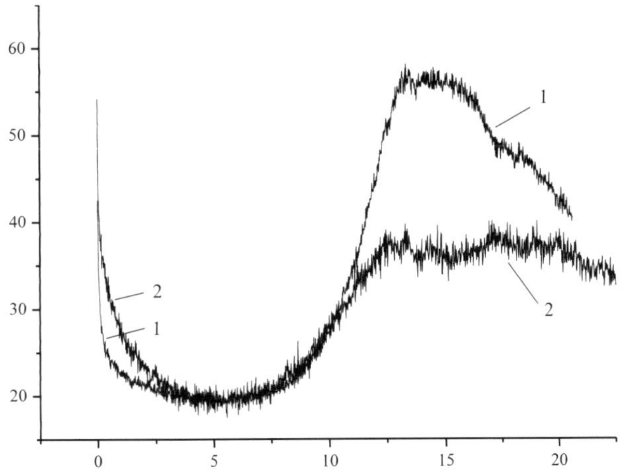


**Figure 1.** Effect of the peptide CLV3 on the superweak photon emission (SWFE), of the planaria *G. tigrina* during the 24 hr after decapitation.
Abscissa is the time of regeneration in hours.
Y-axis is the intensity SWFE in photons per second.
1 – dynamics of SWFE during of regeneration under the action of peptide CLV3 ($10^{-9}$M).
2 – dynamics of SWFE during of regeneration in the control group.

The peptide CLV3 ($10^{-9}$M) stimulated cell proliferation, on average, by 17% compared to the control. Further study of the regeneration of these planar groups at 72 and 120 hr after transection by IDM also revealed the effect of stimulation of planar regeneration by the CLV3 peptide in five series of experiments.

The CLV3 peptide at a concentration of $10^{-9}$M accelerates the regeneration of the planaria by 1.3 times at 72 hr and by 1.2 times after 120 hr after the operation compared to the control samples (Figure 2).

However, unlike the concentration of $10^{-9}$ M, the study of the stimulating effect of the peptide at concentrations of $10^{-12}$ M, $10^{-14}$ M and $10^{-18}$ M did not significantly differentiate the experimental samples from the control ones. Also, the effect of the peptide on the viability of mesenchymal stem cells *Th* from the human tooth pulp was not detected. This result was obtained both in the cultivation of cells without a change in the medium, and in a daily change in the medium and in applying a fresh dose of the peptide during the 3 days of the experiment.

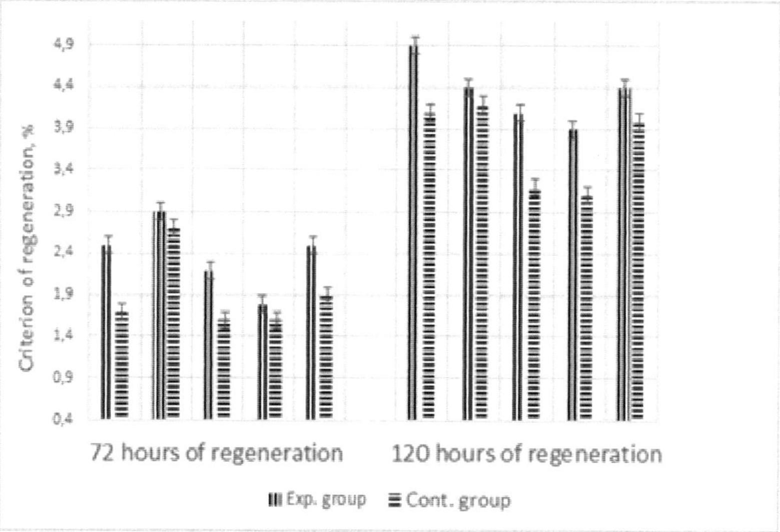


**Figure 2.** The action of the peptide on the regeneration of the planaria on the 3rd and 5th day of regeneration in comparison with the control group. On the y-axis – the regeneration criterion in percent.

Thus, in our experiments, the specificity of direct and cross-reactivity of regulatory peptides on plant, animal and human cells were revealed. Under the action of animal peptides on the morphogenesis of animals and plant peptides on the morphogenesis of plants, effects of superweak concentrations were obtained: up to $10^{-15}$ M for luteinizing hormone releasing hormone (LHRH) [1,3] and up to $10^{-12}$ M for peptide plants pCLV3 [5]. When the plant peptide cross-reacts to the cells of animals and humans, it acts only at a concentration of $10^{-9}$M on the regeneration of the planaria or does not at all affect the survival of human stem cells.

It is interesting to study whether there is the same specificity of 'direct' and 'cross' action for other peptides of animals and plants on other plant and animal test systems.

The molecular mechanisms of biological processes are similar for both plant and animal biosystems. Plants and animals constantly interact with each other at the macro level through symbiosis, parasitism, and mimicry. The same processes take place at the micro level, as well, in particular, at the level of conservative cellular processes, such as proliferation and differentiation of cells, which underlie the growth and morphogenesis of both plants and animals.

However, unified (general) mechanisms of regulation of the morphogenesis of animals and plants at the epigenetic level have not been studied enough. This also applies to the mechanisms of peptide regulation of cell proliferation and differentiation.

An example of such a cross-regulating system is the detection of receptors for plant peptides of the CLE group in parasitic flatworms [14]. Expanding the understanding of the role of this regulatory system for general mechanisms of plant and animal morphogenesis will make it possible to clarify, among other things, the unification and specificity of these molecular instruments as a universal system of peptide regulation in living systems.

**Acknowledgments**
The work was carried out within the framework of the State PushchGENI order, commissioned by the Ministry of Education and Science of the Russian Federation No. 6.10118.2017 / NM for 2017, No. 6.12959.2018 / 12.1 for 2018 and RFBR grant No.16-37-10489, No.14-50-00131 and RFBR ofi_m No. 15-29-01135.

**References**

[1] Sheimann I, Tiras Kh and Balabanova E 1989 *Physiol. J. USSR* **75** 619–25
[2] Tiras Kh, Rubina A, Miloserdov Yu and Tishchenko V 1990 *Morphogenetically active substances* ed I Sheimann (Pushchino) p 134
[3] Sheiman I and Kreshchenko N 2008 *Biomedicine* **1** 79
[4] Kondo T, Sawa S, Kinoshita A, Mizuno S, Kakimoto T, Fukuda H and Sakagami Y. 2006 *Science* **313** 845–8
[5] Konovalov A, Ryzhkina I, Salakhutdinova O, Murtazina L, Shevelev M, Voeikov V, Buravleva E, Glybin A and Skripnikov A. 2017 *Russian Chemical Bulletin* **66** 1699–705
[6] Tiras Kh, Petrova O, Myakisheva S, Deev A and Aslanidi K 2015 *Fundamental Research* **2** 1412–6
[7] Tiras Kh, Gudkov S, Emelyanenko V and Aslanidi K. 2015 *Applied Physics Research* **7** 6
[8] Tiras Kh, Srebnitskaya L, Ilyasova E, Klimov A, Lednev V 1996 *Biophysics* **41** 837–42
[9] Poltavtseva R, Nikonova Yu, Selezneva I, Yaroslavtseva A, Stepanenko V, Esipov R, Pavlovich S, Klimantsev I, Tyutyunnik N, Grebennik T, Nikolaeva A and Sukhikh G 2014 *Exp. Biol. Med* **158** 164–9
[10] Mossman T 1983 *J. Immunol. Methods* **65** 55–63
[11] Pirotte N, Stevens AS, Fraguas S, Plusquin M, Van Roten A, Van Belleghem F, Paesen R, Ameloot M, Cebrià F, Artois T and Smeets K 2015 *Oxid Med Cell Longev* 2015:392476
[12] Ito Y, Nakanomyo I, Motose H, Iwamoto K, Sawa S, Dohmae N and Fukuda H 2006 *Science* **313** 842–5
[13] Wenemoser D and Reddien P 2010 *Developmental Biology* **344** 979–91
[14] Wang X, Mitchum M, Gao B, Li C, Diab H, Baum T. and Davis E 2005 *Molecular Plant Pathology* **6** 187–91

# Effects of *Rhodococcus*-biosurfactants on the molybdenum ion phytotoxicity

**A V Tishchenko[1,2], L V Litvinenko[1] and S A Shumikhin[2]**

[1]Institute of Ecology and Genetics of Microorganisms, Ural Branch of Russian Academy of Sciences, Perm, Russia
[2]Perm State National Research University, Perm, Russia

E-mail: arty.tishchenko@gmail.com

**Abstract.** The effect of molybdenum on the germination of seeds of agricultural plants common vetch, white mustard and oats in the presence of *Rhodoccocus*-biosurfactants was studied. Molybdenum in concentrations from 10 MPC and above had a pronounced inhibitory effect on the germination of seeds. It has been found that pretreatment of seeds and their germination in the presence of biosurfactants contributed to a significant (up to 4.5 times) increase in germination, germinative energy and viability of *Avena sativa* L., *Sinapis alba* L. and *Vicia sativa* L.

## 1. Introduction

The environmental pollution by heavy metals (HMs) is one of the most common and environmentally hazardous for living organisms types of technogenic pollution. The industrial production is accompanied by uncontrolled entry of non-degradable and highly toxic waste containing high concentrations of HMs into natural ecosystems [1].

Molybdenum which is a HM, belongs to 4.1. Hazard Class. Long-term exposure of highly dispersed aerosols of molybdenum and its oxides on the human body causes atrophic catarrh, bronchitis, the initial forms of pneumoconiosis, gastritis, vegetovascular dystonia, and also disrupts metabolism. The effect of high concentrations of molybdenum on higher plants leads to oversaturation and unstable fixation of nitrogen, which in turn leads to the destruction of chlorophyll [2–4].

Traditional ways of soil remediation from HMs are based on the use of physicochemical methods. However, they are not environmentally friendly, since they do not ensure the complete removal of HMs ions from the soil. Modern methods involve the application of biological methods: leaching with the help of surfactants of biogenic origin [5, 6], phytoremediation [7–9]. However, each of these methods is individually ineffective in soil purification. The methods of phytoremediation at high (contamination index > 100 MPC) level of soil contamination with HMs are not used in connection with the death of plants. The use of biosurfactants is limited due to the impossibility of removing desorbed and mobilized forms of HMs from soil. Earlier in *in situ*, we demonstrated the effectiveness of the use of *Rhodoccocus*-biosurfactants to enhance phytoremediation [10]. However, the effect of biosurfactants on plants has not been adequately studied at present.

The aim of this study is to evaluate the effect of molybdenum on germination of seeds and the development of seedlings of green manures in the presence of *Rhodoccocus*-biosurfactants.

## 2. Materials and methods

The seeds of oat (*Avena sativa* L.), white mustard (*Sinapis alba* L.) and common vetch (*Vicia sativa* L.) were used. Molybdenum was used as the salt of ammonium molybdate ($(NH_4)6Mo_7O_{24} \times 4H_2O$) in concentrations multiple of 1, 10, 50, and 100 maximum permissible concentrations (MPC), taking into account the Clark background in terms of pure metal [10]. The concentration of molybdenum corresponded to 10, 100, 500, and 1000 mg/kg of contaminated soil. The phytotoxicity level was determined in accordance with Standard Methodological Recommendations (Phytotest) [12]. Seeds were grown under illumination with white fluorescent lamps, the spectrum of which is as close as possible to daylight; temperature was 23–25°C, pH was 6.3–6.5.

The actinobacterial strain *Rhodococcus ruber* IEGM 231 is active producer of biosurfactants (http://www.iegmcol.ru/strains/rhodoc/ruber/r_ruber231.html). The bacterial culture was grown in RS medium (Rhodococcus Surfactant) on an orbital shaker for 7 days (160 rpm, 28°C). The RS medium contained (g/l) $Na_2HPO_4$ – 2.0; $KH_2PO_4$ – 2.0; $KNO_3$ – 1.0; $(NH_4)_2SO_4$ – 2.0; NaCl – 1.0; $MgSO_4x7H_2O$ – 0.2; $CaCl_2x2H_2O$ – 0.02; $FeCl_3x7H_2O$ – 0.01. *n*-dodecane ($C_{12}$) or *n*-hexadecane ($C_{16}$) were used at a concentration of 3 vol.% as the only sources of carbon and energy. Crude *Rhodococcus*-biosurfactant complexes of glycolipid nature, produced by *Rhodococcus*, were obtained by the method previously described [14]. According this method, 5–7-day bacterial culture grown in RS medium was settled in a separatory funnel, after which the lower transparent aqueous layer was drained. The remaining uniform hydrophobic oily mass was subjected to ultrasonic sounding (30 min) under compulsory cooling conditions. To distinguish the biosurfactants produced by *Rhodococcus* in a liquid mineral medium with *n*-dodecane and *n*-hexadecane we used designations BS $C_{12}$ and BS $C_{16}$, respectively.

Experiments on the germination of seeds were carried out in triplicates. 25 dry and disease-free seeds were placed in Petri dish and 5 ml of emulsion of *Rhodococcus*-biosurfactants and ammonium molybdate was added. As a control, we used: (i) distilled water; (ii) aqueous solutions of biosurfactants in concentrations of 2.0, 4.0 and 8.0 g/l; (iii) aqueous solutions of ammonium molybdate. The germinative energy was determined on the third day, the germination and length of the seedlings were determined on the seventh day of the experiment. Statistical analysis of the obtained results was carried out by standard methods with calculation of the mean, standard deviation and confidence interval using the Microsoft Excel 2007 software package.

## 3. Results and discussion

A positive correlation between the effect of seed treatment by *Rhodococcus*-biosurfactants on the decrease in phytotoxicity of ions molybdenum was revealed. The germinative energy of seeds of all plant species used in the work in the presence of *Rhodococcus*-biosurfactants is 2.0–4.5 times higher than that in conditions of contamination with molybdenum.

Table 1 shows the effect of $Mo^{6+}$ ions on the length of shoots and roots of oat, mustard and vetch propagation in the presence of BS $C_{12}$. According to our data, pretreatment of seeds with this biosurfactant did not have a significant effect on seed growth.

While pretreatment of plant seeds with *Rhodococcus*-biosurfactants, produced in the presence of *n*-hexadecane, increases the shoots and roots of all the plant species studied more intensively (up to 2.5 times) (table 2).

After preliminary treatment of seeds with biosurfactants, their germination capacity increased to 4.5 times. As can be seen from figure 1, the highest germination (100%) of *Avena sativa* L. seeds was detected in the presence of molybdenum in concentrations up to 10 MPC. Apparently, molybdenum stimulates the germination of oat seeds, since it is a vital element. At higher concentrations (50 MPC and higher), molybdenum inhibited the germination process of seeds. Germination of seeds treated with BS $C_{16}$ was 1.2–1.6 times higher than that of BS $C_{12}$.

**Table 1.** Effect of molybdenum on the length of shoots and root system (mm) of oats, mustard and vetch sprouts in the presence of BS $C_{12}$.

| Variants | *Avena sativa* L. | | *Sinapis alba* L. | | *Vicia sativa* L. | |
|---|---|---|---|---|---|---|
| | Length of shoots | Length of roots | Length of shoots | Length of roots | Length of shoots | Length of roots |
| Control | 49.1±8.3 | 40.8±5.2 | 25.2±9.9 | 22.5±4.8 | 27.3±6.0 | 19.3±9.8 |
| $Mo^{6+}$ 1 MPC | 44.2±15.5 | 55.5±19.3 | 23.3±11.3 | 33.0±28.1 | 8.4±4.2 | – |
| $Mo^{6+}$ 10 MPC | 47.2±15.1 | 24.5±11.3 | 11.0±2.7 | 10.2±3.8 | 15.0±2.6 | – |
| $Mo^{6+}$ 50 MPC | 37.9±12.6 | 6.9±3.4 | – | – | – | – |
| $Mo^{6+}$ 100 MPC | 24.8±11.6 | 8.4±3.5 | – | – | – | – |
| BS $C_{12}$ 2 g/l | 55.7±14.4 | 48.4±9.5 | 32.2±8.7 | 36.7±3.0 | 30.8±2.6 | 18.2±3.0 |
| BS $C_{12}$ 4 g/l | 67.4±5.7 | 53.0±6.1 | 39.7±6.7 | 37.0±3.8 | 32.5±1.8 | 19.3±3.4 |
| BS $C_{12}$ 8 g/l | 45.1±5.2 | 35.1±3.2 | 24.6±5.0 | 26.2±8.0 | 27.0±2.0 | 12.3±3.4 |
| 1 MPC / 2 g/l | 60.6±25.9 | 58.0±24.2 | 23.3±9.6 | 18.4±18.8 | – | – |
| 1 MPC / 4 g/l | 64.1±25.0 | 56.5±29.8 | 15.9±6.4 | 33.8±18.7 | – | – |
| 1 MPC / 8 g/l | 51.8±23.1 | 46.7±24.8 | 16.7±9.3 | 21.1±10.7 | – | – |
| 10 MPC / 2 g/l | 54.6±20.0 | 24.5±8.3 | – | – | – | – |
| 10 MPC / 4 g/l | 44.3±16.2 | 28.3±7.0 | – | – | – | – |
| 10 MPC / 8 g/l | 52.1±21.3 | 13.0±11.6 | – | – | – | – |

The solutions used in the work were presented as the concentrations of molybdenum ions / *Rhodococcus*-biosurfactants (MPC / g/l).

**Table 2.** Effect of molybdenum on the length of shoots and root system (mm) of oats, mustard and vetch sprouts in the presence of RB $C_{16}$.

| Variants | *Avena sativa* L. | | *Sinapis alba* L. | | *Vicia sativa* L. | |
|---|---|---|---|---|---|---|
| | Length of shoots | Length of roots | Length of shoots | Length of roots | Length of shoots | Length of roots |
| RB $C_{16}$ 2 g/l | 39.5±26.0 | 41.8±21.3 | 30.5±11.1 | 33.6±26.3 | 29.2±2.8 | 21.3±1.6 |
| RB $C_{16}$ 4 g/l | 31.9±19.3 | 35.3±18.3 | 27.9±8.7 | 33.8±20.1 | 31.8±1.7 | 19.8±0.9 |
| RB $C_{16}$ 8 g/l | 23.5±7.0 | 23.9±13.6 | 20.5±7.4 | 40.2±29.7 | 30.8±1.8 | 25.4±2.2 |
| 1 MPC / 2 g/l | 69.2±24.3 | 72.8±30.1 | 24.2±8.1 | 39.9±20.8 | 17.4±3.6 | 10.3±4.4 |
| 1 MPC / 4 g/l | 57.5±31.4 | 56.6±27.3 | 25.7±8.3 | 39.5±21.0 | 17.0±0.8 | 12.0±3.2 |
| 1 MPC / 8 g/l | 47.3±36.5 | 61.5±28.5 | 33.9±6.8 | 34.2±11.6 | 14.7±2.1 | 13.0±3.6 |
| 10 MPC / 2 g/l | 55.3±22.0 | 29.5±17.3 | 16.7±6.4 | 12.8±3.8 | 24.5±0.8 | 8.6±2.3 |
| 10 MPC / 4 g/l | 56.0±17.2 | 23.5±13.5 | 17.2±7.1 | 15.7±4.7 | 18.0±1.0 | 9.2±0.8 |
| 10 MPC / 8 g/l | 51.5±27.2 | 34.5±15.9 | 20.1±6.3 | 21.4±12.3 | 18.0±1.0 | 10.0±0.2 |
| 50 MPC / 2 g/l | 48.1±5.6 | 17.6±1.8 | 6.3±2.0 | – | 9.3±2.5 | 9.6±5.8 |
| 50 MPC / 4 g/l | 40.8±14.6 | 14.1±8.3 | 6.3±1.2 | – | 8.4±3.6 | 17.0±2.6 |
| 50 MPC / 8 g/l | 42.7±3.5 | 12.4±2.6 | – | – | 8.0±2.2 | 4.5±0.7 |
| 100 MPC / 2 g/l | 29.8±5.9 | 15.0±1.4 | – | – | 5.5±2.1 | 9.0±0 |
| 100 MPC / 4 g/l | 34.0±11.8 | – | – | – | 3.5±1.2 | 6.4±0.5 |
| 100 MPC / 8 g/l | 28.1±4.6 | – | – | – | 7.6±1.5 | 5.4±2.3 |

The solutions used in the work were presented as the concentrations of molybdenum ions / *Rhodococcus*-biosurfactants (MPC / g/l).

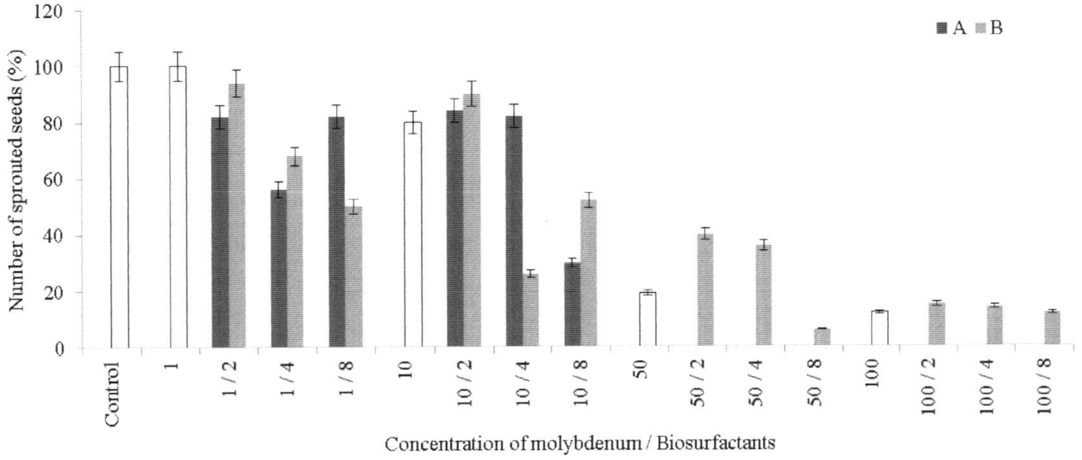


**Figure 1.** Influence of biosurfactants on the germination capacity of *Avena sativa* L. seeds in the presence of molybdenum ions.
A BS $C_{12}$, B – BS $C_{16}$; 1, 10, 50, 100 – concentration of molybdenum 1, 10, 50, 100 MPC; 1/2, 1/4, 1/8 – concentration of molybdenum, MPC / concentration of biosurfactants, g/l.

The revealed regularity was also maintained when germination of mustard seeds: the amount of germinating seeds when processing them BS $C_{16}$ to 3.5 was higher, compared with that after treatment with BS $C_{12}$. Molybdenum at a concentration of 50 MPC inhibited the germination of mustard seeds, whereas after processing with *Rhodococcus*-biosurfactants, up to 15% of mustard seeds germinated in the presence of this level of environmental contamination with molybdenum (see figure 2).

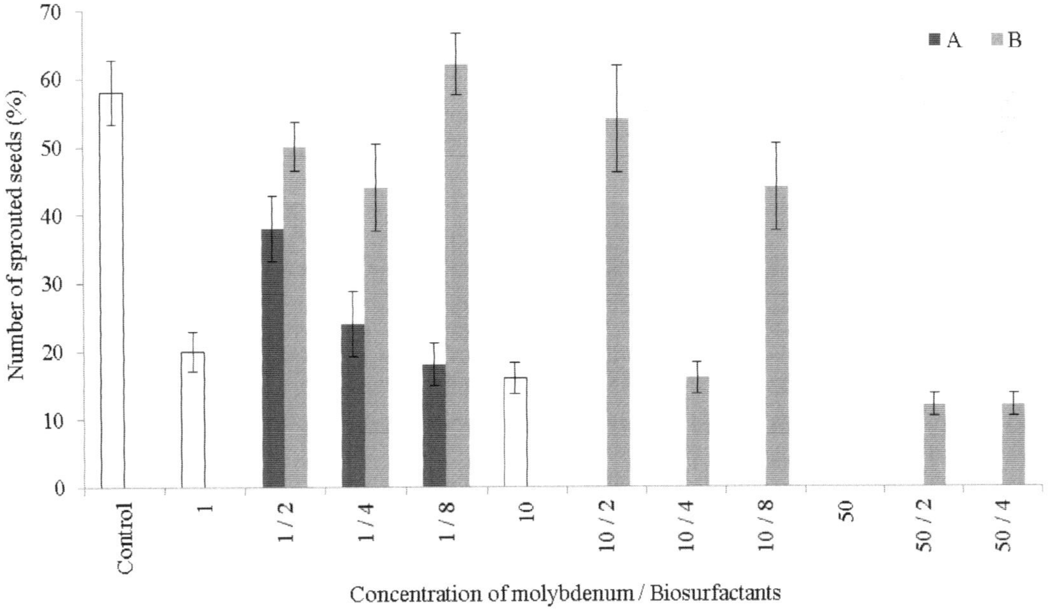


**Figure 2.** Influence of biosurfactants on the germination capacity of *Sinapis alba* L. seeds in the presence of molybdenum ions.
A BS $C_{12}$, B – BS $C_{16}$; 1, 10, 50 – concentration of molybdenum 1, 10, 50 MPC; 1/2, 1/4, 1/8 – concentration of molybdenum, MPC / concentration of biosurfactants, g/l.

A similar dependence was observed in experiments with *Vicia sativa* L. As can be seen in figure 3, the seed germination after treatment with BS $C_{16}$ was repeatedly increased, the vetch seeds germinated even at a molybdenum concentration of up to 100 MPC in the germination medium, whereas BS $C_{12}$ did not significantly affect the germination of the vetch seeds.

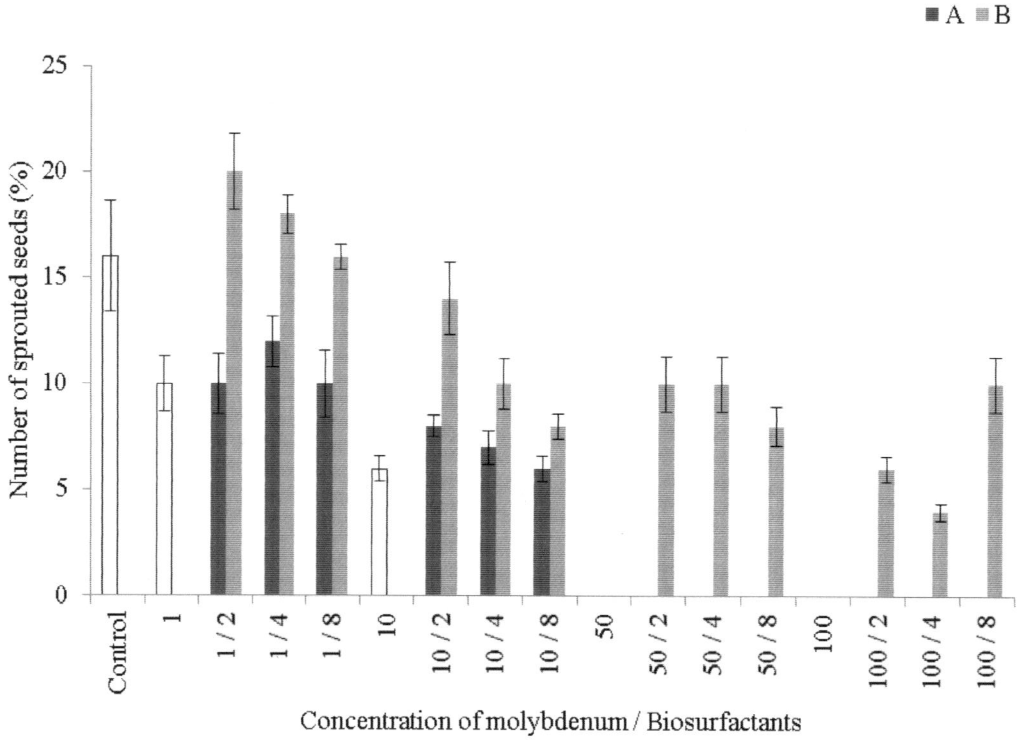


**Figure 3.** Influence of biosurfactants on the germination capacity of *Vicia sativa* L. seeds in the presence of molybdenum ions.
A BS $C_{12}$, B – BS $C_{16}$; 1, 10, 50, 100 – concentration of molybdenum 1, 10, 50, 100 MPC; 1/2, 1/4, 1/8 – concentration of molybdenum, MPC / concentration of biosurfactants, g/l.

## 4. Conclusion

As a result of the conducted studies it was established that the most stable plant to the action of molybdenum was white mustard, less stable plant was oats. In terms of the degree of resistance to molybdenum, the plants used in the work could be distributed in a series: *Sinapis alba* L.> *Vicia sativa* L.> *Avena sativa* L. The dependence of the decrease in phytotoxicity of molybdenum after treatment of BS $C_{16}$ seeds was revealed. Thus, after preliminary treatment of the seeds with an aqueous emulsion of *Rhodococcus*-biosurfactants, the germination and germinative power of the plants studied increased to 4.5 times and the growth rate of the roots and shoots of the seeds under study in conditions of pollution with molybdenum increased up to 2.5 times.

**Acknowledgments**

The work was carried out within the framework of the state assignment registration number of the topic: 01201353247 and Complex program 18-4-8-21.

## References

[1] Khalid S, Shahid M, Niazi N K, Murtaza B, Bibi I and Dumat C 2017 *J. Geochem. Explor.* **182** 247-68

[2] Mohd Adnan A S, Abu Zeid I M, Ahmad S A, Halmi M I E , Abdullah S R S , Masdor N A, Shukor M S and Shukor M Y 2016 *MJSS* **20** 111-34

[3] Shukor M Y and Syed M A 2010 *Current Research, Technology and Education Topics in Applied Microbiology and Microbial Biotechnology* **2** 1304-10

[4] Burzyński M and Buczek J 1989 Interaction between cadmium and molybdenum affecting the chlorophyll content and accumulation of some heavy metals in the second leaf of Cucumis sativus L. *Acta Physiol. Plant.* **11** 137–45

[5] Vijayanand S and Divyashree M 2015 Bioremediation of heavy metals using biosurfactant producing microorganisms *Int J Pharm Sci Res* **6** 840–7

[6] Kostina L V, Kuyukina M S and Ivshina I B 2009 *Bulletin of Perm University. Biology* **10** 95–110

[7] Mishra J, Singh R and Arora N K 2017 Alleviation of heavy metal stress in plants and remediation of soil by Rhizosphere microorganisms *Front Microbiol* **8** 1706–13

[8] Ojuederie O B and Babalola O O 2017 Microbial and plant-assisted bioremediation of heavy metal polluted environments: A review *Int J Environ Res & PH* **14** 1504–11

[9] Mukhtar B, Malik M F, Shah S H, Azzam A, Liaqat S and Liaqat I 2017 *Int. J. of Applied Biology and Forensics* **15** 5–15

[10] Kostina L V, Kuyukina M S and Ivshina I B 2014 *Bulletin of Perm University* **4** 73–8

[11] GN 2.1.7.2041-06. [Soil, purification of populated areas, residuals. Threshold limit concentration of chemical substances in the soil] Available at: http://www.gosthelp.ru/text/GN217204106Predelnodopust.html

[12] MR 2.1.7.2297-07. [Justification of the hazard class of production and consumption waste on the ground of its phytotoxicity] Available at: http://www.ohranatruda.ru/ot_biblio/normativ/data_normativ/52/52957

[13] Ivshina I B, Kuyukina M S, Philp J C and Christofi N 1998 Oil desorption from mineral and organic materials using biosur factant complexes produced byRhodococcus species *World J Microb Biot* **14** 711–17

[14] Kuyukina M S, Ivshina I B, Philp J C, Christofi N, Dunbar S A and Ritchkova M I 2001 Recovery of *Rhodococcus biosurfactants* using methyl-tertiary butyl ether extraction *J Microbiol Meth* **46** 149–56

# Control of pollution of the air of urbanized places at use of pesticides on the basis of triazolinthione

**M S Grechina and A A Ivchenkova**

Federal Scientific Center of Hygiene named after F. F. Erisman of Rospotrebnadzor, Moscow, 141014 Russian Federation

E-mail: mgsea@mail.ru

**Abstract**. An analytical solution to control the active ingredient in the air, a pesticide of the chemical class of triazolinthione – prothioconazole, which is a systemic fungicide of a new generation that has a protective, eradicating and curative effect, the article presents. The method is based on HPLC with an ultraviolet detector (detection wavelength 213 nm), involves the air samples on high-density paper filter with an aspiration rate of 5 l/min. Extraction of prothioconazole from the filters is performed with acetonitrile. To concentrate the extract from the filters, solid-phase extraction using octadecyl silane-based cartridges is used. Was noted that the concentration cannot be performed directly without diluting the aliquot of the extract with water in a ratio of 1:9. Due to the special properties of this active ingredient, its propensity for degradation, the amino acid cysteine is used to stabilize the aqueous solutions obtained. The linearity of the calibration characteristic was confirmed in the concentration range 0.05-0.5 μg/ml (correlation coefficient more than 0.999). The lower limit of the quantitation of prothioconazole in the air environment is 0.0025 mg/m3 when aspirating 80 L of air, which is 8 times lower than the established value of approximate safe level of influence prothioconazole in the air of urbanized places (0.02 mg/m$^3$). The total measurement error doesn't exceed 16%.

## 1. Introduction

Prothioconazole (RS)-2-(2-(1-chlorocyclopropyl)-3-(2-chlorophenyl)-2-hydroxypropyl-2,4-dihydro-1,2,4-triazol-3-thione) belongs to a relatively new group of active substances of the chemical class of triazolinethions, which is open by structural variation of the azole heterocycle [1]. The substance was first obtained from azole by its interaction with butyllithium and then sulfur [2]

Prothioconazole is a systemic fungicide of a new generation that has a protective, eradicating and prolonged curative effect [3]. According to studies conducted by the ISPA (Institute of Science of Food Production, Italy), the use of formulations containing prothioconazole effectively inhibits the development of rust, powdery mildew, septoria, fusarium spike [4], friability of stems and leaf patches on barley, wheat and other crops [5,6]. In this connection, preparations based on prothioconazole, are becoming increasingly common, while being used both individually and in a mixture with fungicides, insecticides and herbicides. At present, 10 formulations based on prothioconazole are allowed on the territory of the Russian Federation.

In the environment, prothioconazole metabolizes to a more stable compound, prothioconazole-desthio [7]. This metabolite influences the formation of powerful shoots, a well-developed root system, increased bushiness, drought resistance, and high-quality grain parameters, and provides reliable protection of the culture against many types of diseases and harmful insects at the initial stages of its growth [8, 9].

The established value of ADI of prothioconazole is 0.05 mg/kg of body weight, approximate safe level of prothioconazole in air of urbanized places is 0.02 mg/m$^3$. The existing official methods for measuring the concentration of prothioconazole in the air provide a detection lower limit of 0.05 mg/m$^3$.

The main aim of current work was to develop a method for measuring the concentrations of prothioconazole in the air of urbanized place using high-performance liquid chromatography with an ultraviolet detector to monitor air pollution when a pesticide is used in agricultural production.

## 2. Materials and Methods

An analytical standard sample of prothioconazole (99.8% main component content); water and acetonitrile qualifications for HPLC, orthophosphoric acid (85%), glacial acetic; methanol, ethanol 95% and L-cysteine was used.

Identification and quantification of prothioconazole was performed on an Agilent 1200 liquid chromatograph equipped with a diode array detector, a binary pump, a vacuum degasser, a thermostatically controlled column compartment and a standard autosampler. Chromatographic steel column 150 mm long, inner diameter 4.6 mm, containing sorbent C18, 5μm granulation (ZORBAX Eclipse XDB-C18 (Cat. No. 993967-902); room temperature; isocratic elution – mobile phase: 0.02 M orthophosphoric acid – acetonitrile (40:60 v/v); the flow rate of the eluent is 1.0 cm$^3$/min; the volume of the injected sample is 30 μl.

The stock solution of prothioconazole with a concentration of 100 μg/ml and with a concentration of 10 μg/ml were prepared in acetonitrile. Working solutions for calibration with concentrations of prothioconazole 0.05; 0.1, 0.2, 0.35 and 0.5 μg/ml were prepared in the mobile phase and used freshly prepared.

High-density paper filters, diameter 7.0 cm were used for sampling; SPE cartridges of Waters Sep Pak C18 Classic (Cat. No. WAT051910) were used for sample preparation.

Sampling of the air is made by air aspiration through a high-density paper filter with a volume flow of 5 l/min. Extraction of the substance from the filters was carried out with acetonitrile (twice in 10 ml portions). Half of the extract from the filter (10 ml) was diluted with water in a ratio of 1:9. 0.02 ml of glacial acetic acid and 0.005 g of L-cysteine were added to the solution, after mixing solution passed through pre-conditioned with 2 ml of methanol, following 5 ml of 0.05% acetic acid SPE Sep Pak C18 cartridge. After applying the sample, the cartridge was washed with 5 cm$^3$ of a mixture of acetonitrile-0.05% acetic acid (2:8, by volume) to remove the interfering components. The substance was eluted from a cartridge 2 ml of the same mixture in a volume ratio of 9:1.

Statistical analysis included the determination of the mean value and the relative standard deviation (RSD) from the results of a study of model air samples with theintroducing of prothioconazole.

## 3. Results

A preliminary evaluation of the aggregate state of prothioconazole in the air was performed using computational methods [8] based on the saturated vapor pressure of the substance at a specific temperature (less than $4\times10^{-7}$ Pa at 20 °C) and molecular weight (344.3). The calculated value of the natural volatility of prothioconazole ($0.6\times10^{-10}$ mg/m$^3$) is significantly lower than the approximate safe level of influence of prothioconazole in the air of urbanized places (0.02 mg/m$^3$), which makes it possible to make a conclusion about the hygienic significance of the presence of the substance in the air of urbanized place as an aerosol. Therefore, when sampling, it is possible to use air aspirating through high-density paper filter, with an increase in the collection time to 16 min and a volumetric flow rate of up to 5 L/min. Acetonitrile was used for extraction.

Several wavelengths of the maximum absorption (213, 230, 280 nm) are observed in the absorption spectrum of prothioconazole in the UV area of a range. At the same time, the greatest response is observed at 213 (which is chosen as the main one) and 280 nm. To improve the reliability of substance identification, the analysis was performed at 2 wavelengths.

The correlation between the injected amount of substance and detector response was linear ranging from 0.05 μg/ml to 0.5 μg/ml (correlation coefficient is 0.99956), signal-to-noise ratio at the detection limit of 23:1. On this figure 1 we can see calibration characteristic in the concentration range 0.05-0.5 $\mu g/cm^3$.

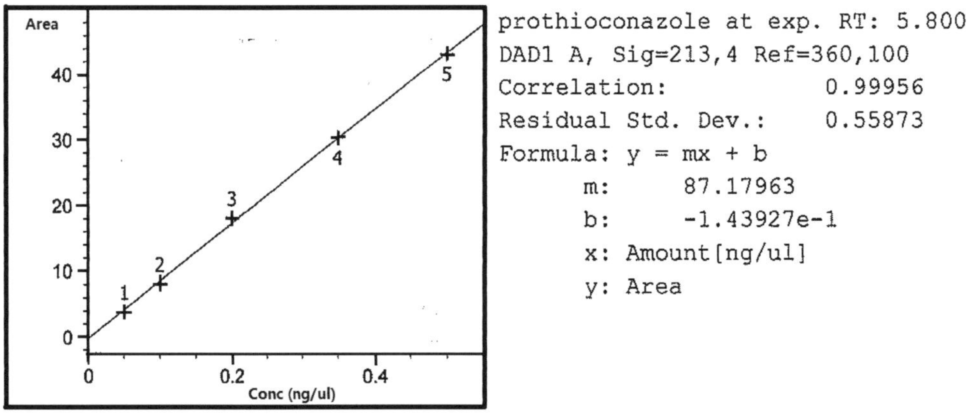


**Figure 1.** Calibration characteristics.

Under the selected chromatographic conditions, prothioconazole forms a clear symmetrical peak at a retention time of 5.7–5.9 minutes. The figure 2 and figure 3 show chromatograms of the calibration solutions with the concentration of prothioconazole 0.05 $\mu g/cm^3$ and 0.05 $\mu g/cm^3$.

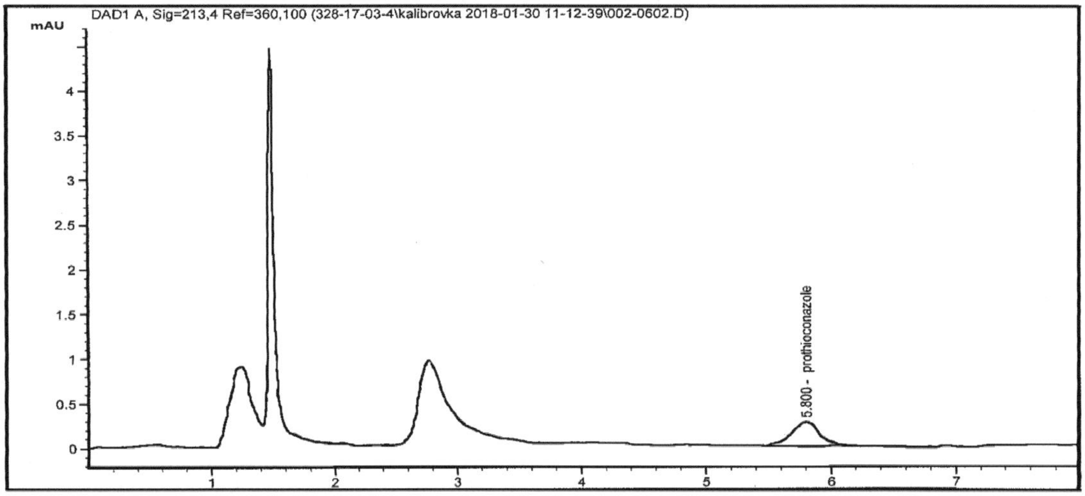


**Figure 2.** Chromatogram of the calibration solution with the concentration of prothioconazole 0.05 $\mu g/cm^3$.

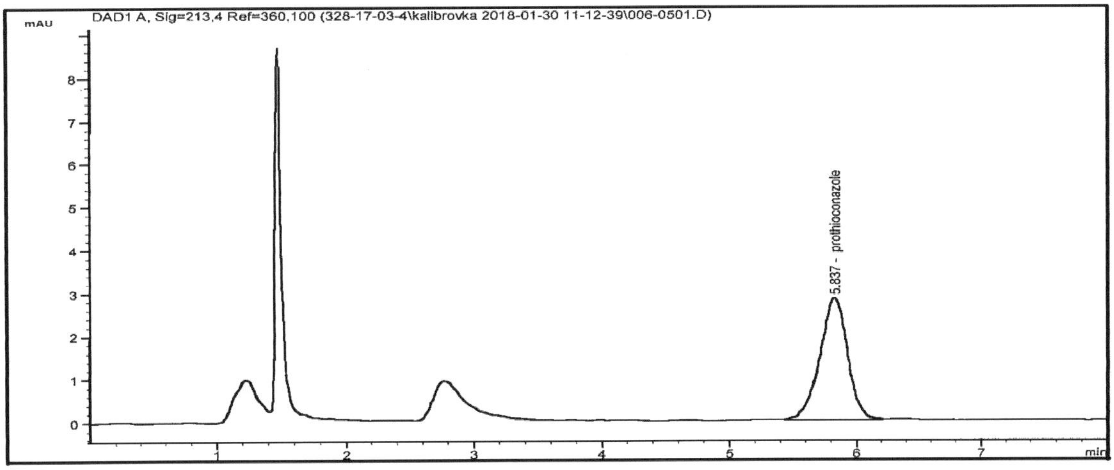


**Figure 3.** Chromatogram of the calibration solution with the concentration of prothioconazole 0.5 μg/cm$^3$.

The lower limit of the quantitation in the injected volume of the sample is 1.5 ng. The lower limit of the quantitation in the air is 0.0025 mg/m$^3$ (the total volume of the sampled air is 80 L). It has been shown that solutions of prothioconazole in acetonitrile with a concentration of 100 μg /ml and 10 μg/ml can be stored in a freezer at a temperature of -18 °C for a month.

The determination of the recovery prothioconazole from the analytical samples was carried out experimentally in 4 levels. To prepare model air samples, to high-density paper filters, the substance was applied over the entire measurement range (0.2–2 μg) using a solution of prothioconazole in acetonitrile at a concentration of 10 μg/ml with varying aliquots. The established range of recovery of prothioconazole from filters is 87.3–111.8% (mean 95.1%). The relative standard deviation of repeatability is 1.77–6.02%. It has been experimentally established that the exposed filters can be stored in the freezer (temperature -15÷-20 °C) for not more than 10 days.

The total error of the procedure for measuring the concentrations of prothioconazole in atmospheric air is 16%.

## 4. Discussion

The development of a method for measuring the concentrations of prothioconazole in the air of urbanized places included justification of the conditions for sampling, sample preparation and subsequent chromatographic analysis. Sampling of the air was carried out by aspiration through a high-density paper filter. With an increase in the collection time of up to 16 minutes and a volumetric flow rate of up to 5 l/min, no breakthrough of the substance was noted.

Prothioconazole is poorly soluble in water, soluble well in ethyl acetate, acetone, acetonitrile, simplest alcohols, weakly in lower limit hydrocarbons, low photolytic resistance to sunlight, average photochemical half-life is 47.7 hours [10].

Providing the necessary sensitivity of the measurement of prothioconazole in atmospheric air, in addition to increasing the volume of sampled air, required the procedure for concentrating the samples. Concentration of the sample by evaporation using a rotary vacuum evaporator and further dissolving the residue in a small volume of the desired solvent is not acceptable for this substance, which is characterized by low stability, is prone to degradation during sample preparation. A known effective method of extracting a substance from a solution is solid-phase extraction, which combines both the concentration and purification functions of the sample. In the framework of this work, Sep Pak C18 concentrating cartridges are used for sample preparation [11]. Satisfactory extraction of the prothioconazole on the sorbent was achieved by diluting an aliquot of the extract with water in a volume ratio of 1:9. For the elution, a mixture of acetonitrile-0.05% acetic acid in a volume ratio of 9:1 was used.

Because of the very high lability of the substance being determined (it is shown that when the water samples are stored at a temperature below -18 °C, losses of up to 15% are noted), in the preparation of samples, the stabilization of the prothioconazole by introducing into the aqueous sample a new component, the amino acid L-cysteine $CH_2(SH)$-$CH(NH_2)$-COOH.

The molecular structure of prothioconazole involves the formation of two competitive tautomeric forms: thion ↔ thiol. Figure 4 shows two tautomeric forms of prothioconazole.

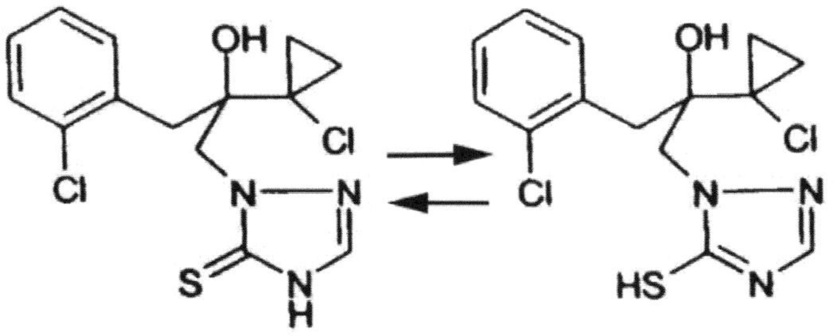

**Figure 4.** Tautomeric forms of prothioconazole: thion ↔ thiol.

When the equilibrium is shifted to the right, the predominant formation of thiol occurs. Further degradation to -S-CH3 or oxidation to the sulfonic acid -SO3H, followed by cleavage of the group, is accompanied by isolation of the prothioconazole-desthio. To achieve a shift of chemical equilibrium to the left, i.e. in the direction of the predominance of the original form of the active ingredient, the amino acid L-cysteine containing the thiol group -SH (based on 50 mg of the substance per 1 L solution) was introduced into the solution as a stabilizer. The use of cysteine for the chemical stabilization of various compounds and complexes (flavoring additives, pigmentary dyes, enzymes, nanoproducts) has been widely described in the literature [12, 13].

Due to the high photo lability of the prothioconazole, all sample preparation and analysis operations were performed with the exception of direct sunlight.

The created method was used in studies on the evaluation of exposure levels of a substance in the air in conditions of full-scale application of preparations based on proteoconazole under various agricultural applications: spraying of field crops, mechanized work 3 days after treatment, seed dressing (cereals, soybeans, corn) and planting (potato) material with subsequent or simultaneous planting, treatment of sunflower. The required active ingredient is not identified (less than the lower limit of quantification 0.0025 $mg/m^3$).

## 5. Conclusions

The conditions for the determination of prothioconazole in the air of urbanized places, including sampling for a high-density paper filter, extraction of the substance with acetonitrile, concentration of an aliquot of the extract from the filter, previously diluted with water in a volume ratio of 1:9, on Sep Pak C18 cartridges, are justified. To stabilize the substance in an aqueous solution, the amino acid L-cysteine was used. The developed method provides a lower limit of quantitative measurement of 0.0025 $mg/m^3$ (for aspirating 80 L of air), which is 8 times lower than the established value of the approximate safe level of influence of prothioconazole in the air of urbanized places (0.02 $mg/m^3$). The total error of the procedure for measuring the concentrations of prothioconazole in atmospheric air is not higher 16%.

**References**

[1] Ohkawa H, Miyagawa H and Lee P 2007 *Pesticide Chemistry. Crop Protection, Public Health, Environmental Safety* (WILEY-VCH Verlag GmbH Co.KGaA, Weinheim)
[2] Turner J A 2015 *The Pesticide Manual. 17th Edition* (Alton BCPS)
[3] Haidukowski M, Visconti A, Perrone G, Vanadia S, Covarelli L, Balestrazzi R and Pascale M 2012 *Phytopathologia Mediterranea* **51** 236–46
[4] Treikale O, Afanasieva I and Pugacheva E 2011 *Plant protection and quarantine* **6** 49–50
[5] Baybakova Ye V, Nefedieva Ye E and Khokhlova T V 2017 *Subtropical and ornamental gardening* **61** 138–41
[6] Parker J E, Warrilow A G, Cools H J, Martel C M, Nes W D, Fraaije B A, Lucas J A, Kelly D E and Kelly S L 2011 *Appl Environ Microbiol.* **77** 1460–5
[7] Haas M.and Justus K 2004 *Pflanzenschutz-Nachr.Bayer-English edition* **57** 207–24
[8] Baybakova E V, Nefedyeva E E and Belopukhov S L 2016 *Applied chemistry and biotechnology* **6** 57–64
[9] Parker J E, Cools H J, Fraaije B A, Lucas J.A., Griffiths W J, Kelly D E, Kelly S L, Warrilow A G and Rigdova K 2013 *Appl Environ Microbiol.* **79** 1639–45
[10] Hellpointner E and Borchers H 2004 *Pflanzenschutz-Nachr.Bayer-English edition* **57** 163–80
[11] Deineka V I, Deineka L A, Sidorov A N, Saenko I I and Kostenko M O 2016 *Sorption and chromatographic processes* **16** 624–30
[12] Weerawatanakorn M, Wu J-Ch, Pan M-Hs and Ho Ch-T 2015 *J. Food and Drug Anal.* **23** 176–90
[13] Cortez R, Luna-Vital D A, Margulis D and Gonzalez de Mejia E 2017 *Comprehensive reviews in Food Science and Food Safety* **16** 180–98

IOP Conf. Series: Materials Science and Engineering **487** (2019) 012023 doi:10.1088/1757-899X/487/1/012023

# Antioxidant and genoprotective properties of the antidiabetic drug metformin under X-ray irradiation

**E E Karmanova, S A Abdullaev, V E Ivanov, G M Minkabirova and V I Bruskov**

Institute of Theoretical and Experimental Biophysics, Russian Academy of Sciences, Pushchino, Moscow Region, 142290, Russian Federation

Email: bruskov_vi@rambler.ru

**Abstract**. The antioxidant and genoprotective properties of the antidiabetic drug metformin have been investigated. The research was performed both in vitro and in vivo under X-ray irradiation. It was shown that metformin reduced the formation of hydroxyl radicals and hydrogen peroxide in aqueous solutions irradiated with X-rays. The particular features of its influence on the formation of $H_2O_2$ were revealed. It has been established that metformin reduced the formation of 8-oxoguanine, a key biomarker of oxidative damage to DNA, in vitro. The effect of metformin on the frequency of micronucleus formation in polychromatic erythrocytes of bone marrow of irradiated mice was investigated. Its genoprotective and radiomimitatory properties at different concentrations and times of application are shown.

## 1. Introduction

Sources of radioactive radiation are widely used in various branches of human activity. Therefore, it is urgent to study the physico-chemical mechanisms of their damage to biological systems and the possibilities of neutralizing their harmful effects on living organisms. A number of effective radioprotectors are now known and well-studied. They exhibit protective properties when administered shortly before exposure to radiation [1, 2]. However, the results of their radioprotective actions strongly depend on time of administration prior to irradiation, which largely limits the possibilities of their practical use. Many of these drugs have side actions, and the long-term effects of their use are still unclear [3]. Another class of radioprotective compounds is radiotrimitators. These drugs are used after irradiation and have a therapeutic effect. Radiomitigators are applied to compensate and repair damage caused by ionizing radiation [4]. In medicine, a lot of drugs are actively used to treat various diseases. A new and actual area of prevention and treatment of different forms of radiation damage is the search for effective radiomitigatory agents among them. This approach does not require the study of their toxicity when used in the therapeutic concentrations for treatment. In this paper, some mechanisms of the radioprotective action of the drug metformin have been investigated.

Metformin (1,1-dimethylbiguanidine hydrochloride) is widely used for treatment of hyperglycemia in patients with type two diabetes. According to numerous published data, metformin has a multifactorial effect on the living organism [5]. It is able to modulate the cellular metabolism and production of mitochondrial reactive oxygen species (ROS) by activation of AMP protein kinase (AMPK) [6, 7]. It has also geroprotective properties [8, 9] and anticarcinogenic activity [10, 11]. There are some data on the radioprotective properties of metformin obtained in animals [12] and cell cultures [13]. Although metformin has been used for many years in medicine and in fundamental research, the mechanisms of its multiple actions are yet little understood and contradictory [14]. In this paper, the antioxidant and genoprotective properties of metformin have been studied in vivo and in vitro at X-ray irradiation.

## 2. Methods

### *2.1. X-ray irradiation*
The animals and solutions were irradiated at the Common Use Centre – Group of Radiation Sources of the Institute of Cell Biophysics, Russian Academy of Sciences, using an RUT-15 therapeutical X-ray device (MosRentgen, Russia) at a dose rate of 1 Gy/min (200 kV, 20 mA, focal distance 37.5 cm). Mice (5 animals per group) were irradiated at a dose of 1.5 or 2 Gy.

### *2.2. Measurement of Hydrogen Peroxide*
The concentration of hydrogen peroxide in irradiated solutions of metformin (Sigma-Aldrich, USA) was measured by the method of enhanced chemiluminescence using a luminol–4-iodophenol–peroxidase system [15]. A counter for liquid scintillation Beta-1 (MedApparatura, Ukraine) operating in the mode of single photon counting was used as a highly sensitive chemiluminometer [16]. The high sensitivity of this method allows registration of hydrogen peroxide at concentrations as low as 1 nM. The content of $H_2O_2$ was determined by using the calibration curves of chemiluminescence dependence on the known concentration of $H_2O_2$ in the solution. The concentration of hydrogen peroxide used for calibration was quantitatively estimated by spectrophotometry at 240 nm using a molar absorption coefficient of 43.6 $M^{-1}\times cm^{-1}$. All used reagents were Sigma-Aldrich, USA. Freshly prepared bidistilled water used for dilution had pH of 5.6 and conductivity of 120 μS/m.

### *2.3. Measurement of Hydroxyl Radicals*
The concentrations of hydroxyl radicals were determined using coumarin-3-carboxylic acid (CCA), a highly specific to OH-radicals fluorescent probe [17]. The CCA (Sigma-Aldrich, USA) concentration used in the test solutions was 0.5 mM. The intensity of fluorescence was measured on a Cary Eclipse spectrofluorimeter (Varian, Australia) with $\lambda_{ex}$ = 400 nm and $\lambda_{em}$ = 450 nm. Calibration for formation of hydroxyl radicals was done using a commercial preparation of 7-OH-CCA (Sigma-Aldrich, USA) [18]. 240 nM/Gy was used as standard for the radiation-chemical yield of OH radicals [19]. The measurements were carried out in the Common Use Centre of the Institute of Theoretical and Experimental Biophysics, Russian Academy of Sciences.

### *2.4. Immunoenzyme assay for determination of 8-oxoguanine (8-oxoG) in DNA*
The method used for immunoenzyme assay and properties of 8-oxoG-specific antibodies have been described in detail previously [16, 20]. DNA from salmon sperm (Sigma-Aldrich, USA) was used at a concentration of 400 μg/ml. Metformin was added to DNA solution immediately before the irradiation. The DNA samples were denatured in a water bath for 9 min and cooled on ice. Aliquots of 50 μl were transferred to the wells of immunoassay plates (Costar, USA). DNA was immobilized by a simple dry adsorption procedure with incubation for 3 h at 80°C until the solution was fully evaporated [21]. The blocking of nonspecific sites was performed using 300 μl of solution containing 1% skimmed milk powder (Sampo, Russia) in 0.15 M Tris-HCl buffer, pH 8.5, and 0.15 M NaCl. Then the plates were incubated on a plate shaker for 2 h at 37°C. Formation of antigen complexes anti-mouse IgG–peroxidase conjugate with 8-oxoG-specific antibodies (50 μl/well, 1:1000 (Bio-Rad, USA)) was carried out in the blocking solution by incubation with stirring for 2 h at 37°C. The samples were washed three times (300 μl/well) with a solution of 50 mM Tris-HCl buffer, pH 8.5, and 0.15 M NaCl on a shaker for 3 minutes. After that, a chromogenic substrate (18.2 mM ABTS (Sigma-Aldrich, USA)) and 10 mM hydrogen peroxide in 75 mM citrate buffer, pH 4.2, were added. The optical density of the samples was measured with a microwell plate reader Multiscan FC (Thermo

Fisher Scientific, Finland) at 405 nm. The details of the calibration procedure and determination of the concentration of DNA solutions have been published elsewhere [20, 22].

### *2.5. Micronucleus (MN) test*

The method for determination of micronuclei was described in detail previously [15]. Histological samples for MN-test were prepared as in [23], with minor modifications. Males of random-bred white Kv:SHK mice aged 5-6 weeks weighing 25–29 g (nursery Kryukovo, Russian Academy of Medical Sciences) were used. They were maintained under controlled conditions of temperature (23±2 °C) and were given standard commercial mouse feed (Arno, Russia) and drinking water ad libitum. More than 2000 polychromatophilic erythrocytes (PCE) per mouse were counted. Isotonic metformin solutions (Sigma-Aldrich, USA) were injected intraperitoneally (i.p.) to mice 15 min after the irradiation with a dose of 1.5 Gy. In another series of experiments, mice were irradiated at a dose of 2 Gy and suspension of tableted metformin (Merck Serono, France) was administered orally at 300 mg/kg before or after irradiation. The mice were killed by cervical dislocation 28 h after irradiation. Animals handling were done according to institutional guidelines for animal care. Experiments with mice were approved by the bioethics committee of ITEB RAS.

## 3. The Results and Discussion

### *3.1. The antioxidant properties of metformin*

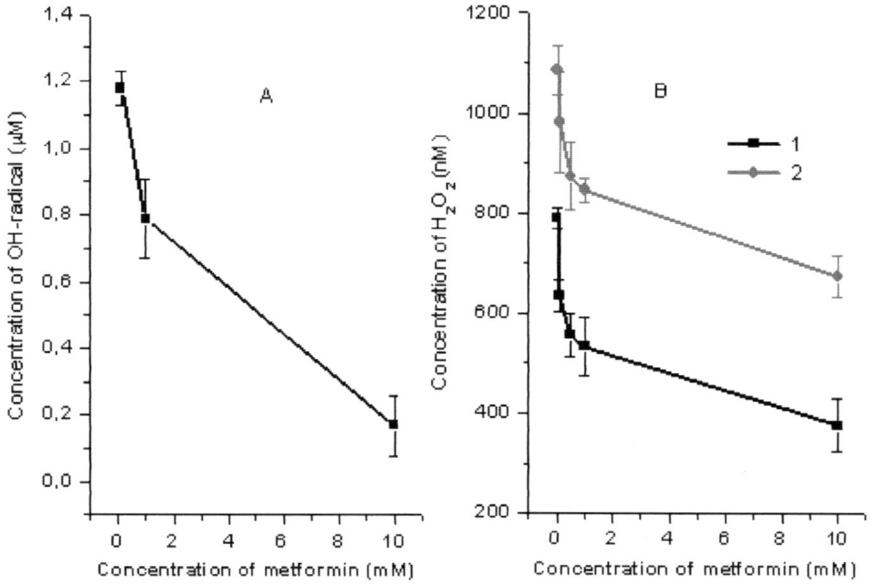


**Figure 1.** (A) Effect of metformin on the generation of hydroxyl radicals in phosphate buffer (1 mM, pH 7.4) irradiated with 5 Gy of X-rays. (B) Effect of metformin on the formation of hydrogen peroxide in phosphate buffer (1 mM, pH 7.4) (1) and in physiological saline (150 mM NaCl) containing phosphate buffer (1 mM, pH 7.4) (2) under the influence of X-ray radiation at a dose of 10 Gy. The background values of hydroxyl radicals and hydrogen peroxide are subtracted from the results. The data are the mean ± SEM (n = 10).

The initial target for the biological impact of ionizing radiation is the aqueous matrix of the organisms with the formation of reactive oxygen species (ROS) as a result of radiolysis of water. Among the ROS the most significant are hydrogen peroxide, as the longest-lived ROS, and hydroxyl radical, as the most reactive. The manifestation of the antioxidant properties of metformin was studied by the formation of hydroxyl radicals and hydrogen peroxide during the radiolysis of 1 mM phosphate buffer, pH 7.4, at X-ray irradiation. It is shown that metformin at a concentration of 10 mM reduces by 95% the formation of hydroxyl radicals under the influence of X-ray radiation at a dose of 5 Gy (figure 1A). Metformin at a concentration of 1 mM and 5 mM reduces the formation of hydrogen peroxide by X-ray irradiation at a dose of 10 Gy by 35% and 55%, respectively. In physiological saline solution, 35% more hydrogen peroxide was formed than in the control sample of 1 mM phosphate buffer, pH 7.4 (figure 1B). The obtained data indicate that metformin is an effective antioxidant serving as scavenger of hydroxyl radicals, whereas formation of hydrogen peroxide can occur not as a result of the recombination of hydroxyl radicals, but largely as a result of the dismutation of hydroperoxide radicals [16].

### *3.2. The genoprotective properties of metformin*

Oxidative damage to DNA is responsible for the genotoxic effects of ionizing radiation. They lead to such biological processes as mutagenesis, carcinogenesis, aging and age-related degenerative diseases. In this work, the genoprotective properties of metformin were investigated in vitro and in vivo. The most sensitive biomarker of oxidative damage to DNA in vitro - the formation of 8-oxoG - was used to determine the genoprotective properties of metformin. It is found that metformin at a concentration of 250 μM in solution of the salmon sperm DNA exposed to X-ray radiation at a dose of 10 Gy reduced the formation of 8-oxoG by 66%, showing genoprotective properties (figure 2).

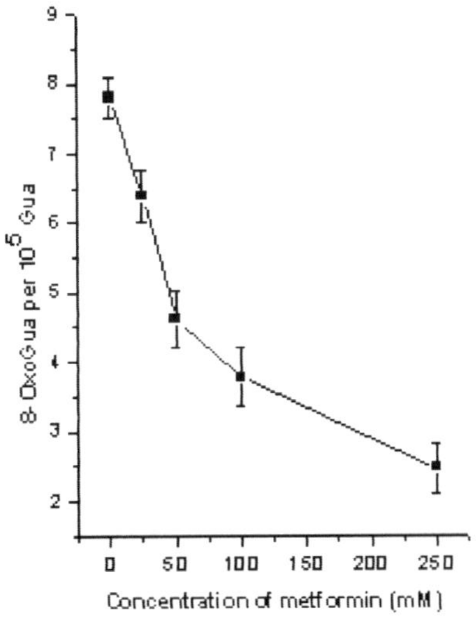


**Figure 2.** Influence of metformin on the formation of 8-oxoguanine in the solution of DNA of salmon sperm in phosphate buffer (1 mM pH 7.4) in vitro under X-ray-irradiation at a dose of 10 Gy. The data are the mean ± SEM (n = 10).

The results of the MN-test showed that metformin, when administered orally at 300 mg/kg, exhibited an obvious genoprotective effect during 6 h with a maximum protection at 15 min after irradiation (figure 3, insertion). Recently it was shown that the same treatment of mice by metformin prolongs their survival rate [24]. Further, the concentration dependence of genoprotective effects of metformin was studied by the MN-test, when metformin was administered to mice at 15 min after irradiation (figure 3). It is shown that metformin (3 mg/kg and 30 mg/kg) administered i.p. after X-ray irradiation at a dose of 1.5 Gy reduces the frequency of formation of PCE with MN by 65 and 75%, respectively. It follows from these data that intraperitoneal injection of metformin at a dose of 30 mg/kg is equivalent to oral administration of metformin at 300 mg/kg.

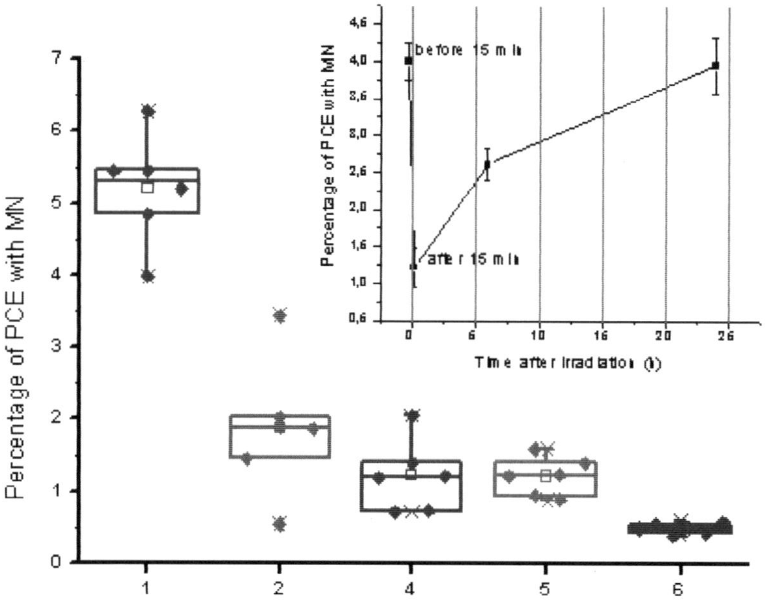


**Figure 3.** Influence of metformin on the frequency of micronucleated polychromatic erythrocytes of bone marrow of X-ray irradiated mice: *1 – 1.5 Gy, 2 – 1.5 Gy and metformin intraperitoneally 3 mg/kg, 3 – 1.5 Gy and metformin intraperitoneally 30 mg/kg, 3 – 2 Gy and metformin orally 300 mg/kg, 4 – intact control 0 Gy*. Insert: Dependence of the frequency of micronucleus formation in polychromatic erythrocytes of bone marrow of mice irradiated at a dose of 2 Gy on time of administration of metformin (per os, 300 mg/kg). The boxplot (or box-and-whiskers plot) consists of a box from the lower quartile of the $x_i$ to their upper quartile, with a crossbar at the median of the $x_i$. Outside of the box, the upper fence is given by $Q_2 + 4(Q_3 - Q_2)$ and the lower fence by $Q_2 + 4(Q_1 - Q_2)$, where $Q_j$ is the $j$th quartile hence $Q_2$ is the median [25]. The whiskers are the horizontal lines going from the box to the most extreme values inside the fences. The dots indicate the position of the data.

According to Stone et al., all the agents that are able to correct radiation injuries can be classified into protectors, mitigators, and treatments depending on the time of their administration [2]. The obtained data demonstrate the possible radiotherapeutic effect of metformin given orally and intraperitoneally immediately and shortly after irradiation. Thus, the antidiabetic drug metformin shows antioxidant, genoprotective and radiomitigating properties under X-ray irradiation and is a promising compound for further studies of its radioprotective potential.

## References

[1] Weiss J F and Landauer M R 2009 History and development of radiation-protective agents *Int. J. Radiat. Biol.* **85** 539–73
[2] Stone H B et al. 2004 Models for evaluating agents intended for the prophylaxis, mitigation and treatment of radiation injuries: Report of an NCI Workshop, 3–4 December 2003 *Radiat. Res.* **162** 711–28
[3] Hosseinimehr S J 2007 Foundation review: Trends in the development of radioprotective agents *Drug Disc. Today* **12** 794–805
[4] Greenberger S, Clump D, Kagan V, Bayir H, Lazo J S, Wipf P, Li S, Gao X and Epperly M W 2012 Strategies for discovery of small molecule radiation protectors and radiation mitigators *Frontiers in oncology* **1** 59
[5] Xu G et al 2015 Metformin ameliorates ionizing irradiation-induced long-term hematopoietic

stem cell injury in mice *Free Radical Biology and Medicine* **87** 15–25

[6] Zhou G et al 2001 Role of AMP-activated protein kinase in mechanism of metformin action *The Journal of clinical investigation* **108** 1167–74

[7] Toyama E Q, Herzig S, Courchet J, Lewis T L, Losón O C, Hellberg K and Shaw R J 2016 AMP-activated protein kinase mediates mitochondrial fission in response to energy stress *Science* **351** 275–81

[8] Anisimov V N 2013 Metformin. Do we finally have an anti-aging drug? *Cell cycle* **12** 3483–9

[9] Cabreiro F et al 2013 Metformin retards aging in C. elegans by altering microbial folate and methionine metabolism *Cell* **153** 228–39

[10] Lee H, Park H J, Park C S, Oh E T, Choi B H, Williams B, Lee C K and Song C W 2014 Response of breast cancer cells and cancer stem cells to metformin and hyperthermia alone or combined *PloS One* **9** e87979

[11] Koritzinsky M 2015 Metformin: a novel biological modifier of tumor response to radiation therapy *Int. J. of Radiat. Oncology• Biology• Physics* **93** 454–64

[12] Miller R C, Murley J S and Grdina D J 2014 Metformin exhibits radiation countermeasures efficacy when used alone or in combination with sulfhydryl containing drugs *Radiat. Res* **181** 464–70

[13] Muaddi H, Chowdhury S, Vellanki R, Zamiara P and Koritzinsky M 2013 Contributions of AMPK and p53 dependent signaling to radiation response in the presence of metformin *Radiotherapy and Oncology* **108** 446–50

[14] Pietrocola F and Kroemer G 2017 Metformin: a metabolic modulator *Oncotarget* **8** 9017

[15] Asadullina N R, Usacheva A M, Smirnova V S and Gudkov S V 2010 Antioxidative and radiation modulating properties of guanosine-5′-monophosphate *Nucleosides Nucleotides Nucleic Acids* **29** 786–99

[16] Bruskov V I, Malakhova L V, Masalimov Zh K and Chernikov A V 2002 Heat-induced formation of reactive oxygen species and 8-oxoguanine, a biomarker of damage to DNA *Nucleic Acids Res* **30** 1354–63

[17] Manevich Y, Held K D and Biaglow J E 1997 Coumarin-3-carboxylic acid as a detector for hydroxyl radicals generated chemically and by gamma radiation *Radiat. Res* **148** 580–91

[18] Kachur A V, Manevich Y and Biaglow J E 1997 Effect of purine nucleoside phosphates on OH-radical generation by reaction of $Fe^{2+}$ with oxygen *Free Radic. Res* **26** 399–408

[19] Ward J F 1988 DNA damage produced by ionizing radiation in mammalian cells: identities, mechanisms of formation and reparability *Progr. Nucl. Acid Res. Mol. Bol.* **35** 95–125

[20] Smirnova V S, Gudkov S V, Shtarkman I N, Chernikov A V and Bruskov V I 2005 The genotoxic action of uranyl ions on DNA in vitro caused by the generation of reactive oxygen species *Biofizika* **50** 456–63

[21] Gudkov S V, Shtarkman I N, Chernikov A V, Usacheva A M and Bruskov V I 2007 Guanosine and inosine (riboxin) eliminate the long-lived protein radicals induced X-ray radiation *Dokl. Biochem. Biophys.* **413** 50–3

[22] Shtarkman I N, Gudkov S V, Chernikov A V and Bruskov V I 2008 Effect of amino acids on X-ray-induced hydrogen peroxide and hydroxyl radical formation in water and 8-oxoguanine in DNA *Biochemistry-Moscow* **73** 470–8

[23] Karp O E, Gudkov S V, Garmash S A, Shtarkman I N, Chernikov A V and Bruskov V I 2010 Genotoxic effect of long-lived protein radicals in vivo generated by X-ray irradiation *Dokl. Biochem. Biophys* **434** 250–3

[24] Abdullaev S, Minkabirova G, Karmanova E, Bruskov V and Gaziev A 2018 Metformin prolongs survival rate in mice and causes increased excretion of cell-free DNA in the urine of X-irradiated rats *Mutation Research/Genetic Toxicology and Environmental Mutagenesis* **831** 13–18

[25] Rousseeuw P J, Ruts I and Tukey J W 1999 The bagplot: a bivariate boxplot. *The American Statistician* **53** 382–7

# Radiation stability of lindane and active substance of hexachlorane dust preparation under electron irradiation

**T V Melnikova[1], L P Polyakova [1], A V Gordeev [2] and A A Oudalova [1,3]**

[1] Obninsk Institute for Nuclear Power Engineering, National Research Nuclear University "MEPhI", Obninsk, Russia
[2] State Research Center – Burnasyan Federal Medical Biophysical Center of FMBA, Moscow, Russia
[3] Russian Institute of Radiology and Agroecology, Obninsk, Russia

E-mail: tritel2010@gmail.com

**Abstract.** Electron beam impact on the radiation resistance of organochlorine hydrocarbon is considered for two forms of pesticide – a pure chemical compound and a component of active substance of a pesticide product to be disposed of. State standard reference samples (SRS) of hexachlorocyclohexane (HCH) – lindane and hexachlorane dust – are used. The fullness of decomposition are determined for lindane and organochlorine compounds as part of the active substance, depending on the aggregate state of the preparation and the absorbed dose of electronic radiation. In all samples, a reduction in the radiation resistance of HCH and its isomers is detected within the range of the radiation dose (D) from 10 to 200 kGy. The maximum decomposition is observed in the suspension of the preparation. The dependence of a degree of decomposition on the amount of radiation dose is not linear.

## 1. Introduction

The disposal of acutely toxic pesticides from the list of persistent organic pollutants (POP) is a priority in the field of hazardous waste management both in the Russian Federation and abroad. Full destruction of accumulated stock of organochlorine pesticides is an important task. Tens of thousands tons of pesticides were found following the inventory audit conducted in Russia in accordance with the international treaty on the disposal of POP [1].

Possibilities of using different types of radiation in the destructive processes for bulk hydrocarbon raw stock are discussed in some studies [2–3], that can include natural gas, gas condensates, oil and oil products etc. The examples of successful implementation of such scientific developments in different economy sectors are provided. Thus, the development of radiation technologies (RT) in the field of large-tonnage POP wastes management can be considered as promising technique [2–3].

A possibility of the POP disposal by means of radiation exposure is primarily connected with the selection of a radiation source. As it follows from the literature data [4] as well as from our primary studies [5], the equal absorbed doses of radiation cause an equivalent destruction effect on organochlorine hydrocarbon in case of both gamma and electron beam impact. From practical point of view, a difference in the degree of decomposition of, e.g., organochlorine pesticides (OCP), which belongs to the POP group, is not so important as compared to the disadvantages and advantages of the process technology. It is noted [4] that in case of gamma-radiation, it is possible to use containers for simultaneous exposure of the substance in large volumes. However, the exposure time for a gamma installation is not less than several hours, while using a high-capacity electron beam accelerator ensures a higher mobility.

In addition to the serch for the best technological solution for the POP radiation disposal, it is important to design the optimal conditions for the radiation exposure of the subject. This study deals with some methodological aspects of electronic impact on the samples including SRS of lindane [6], hexachlorane dust wastes produced in accordance with the technical specifications [7] and the preparation's alkaline water suspension prepared by the method described in [8].

In this work, the dose dependence is studied for the radiation-induced destruction of γ-HCH isomer as a component of an active substance of the pesticide preparation exposed to radiation in different aggregate states.

## 2. Materials and methods

The 'Hexachloran dust' pesticide formulation, intended for utilization, was substituted for radiation exposure. Electronic irradiation was carried out in doses of 10, 44, 100, 120, 150 and 200 kGy given by the radiation-technological unit (RTU) with an UELR-10-10-40 accelerator. At each radiation dose, several samples (3-7) were irradiated; each sample was arranged as a container (test-tube or weighing bottle) with pesticide formulation in a given form. The test-tubes are made of glass, in the form of a cylinder of 17 cm in length and 1.5 cm in diameter (with a stopper), intended for chemical, biological and microbiological laboratory procedures. Irradiation was delivered via multiple transmission of a transport container (box) with samples, placed on the conveyor belt, through the electron-beam; at this, the necessary dose value was supplemented step-by-step. As soon as the required dose was reached in a certain cycle by a set of samples, this set was removed from the transport container. During irradiation process, dose measurements were carried out.

The initial and irradiated samples of the 'Hexachloran dust' pesticide formulation were analyzed for the content of three stereoisomers of the active substance – α-, *β*-, γ-HCH. The active substance was extracted with hexane. Methods of sample preparation for irradiation are presented earlier [8, 9]. The experiment on sample irradiation and subsequent sample preparation for chemical analysis was carried out in triplicate. Samples analysis was carried out on GC Crystal-5000.1 chromatographic installation (Russia) in the column temperature programming mode. Statistical processing of data was done with MS Excel 2016 and Origin 2015.

More detailed information on radiation conditions, characteristics of the radiation device, the methods for analysing and processing the experimental data obtained are described earlier [8, 9]. In this publication, some results of the study are discussed that are important for the variation of technical parameters if electronic radiation sources are used in the RT developed for the above purposes.

## 3. Results and discussion

The radiation destruction of γ-HCH is calculated through the change in the degradation degree of substance (P, %) as part of the preparation's dry powder [9]. The destruction turned out to be instable following the electronic radiation within the range of increasing doses. The substance decomposition efficiency was various for different absorbed doses ranges (0–10, 10–100, 100–200 kGy).

Table 1 presents the results from the degradation degree determination (P, %) for γ-HCH in the samples of lindane and the preparation's suspension analysed, as well as the data for dust radiation degradation [9] in the broad range of doses from electron beam exposure.

According to the results obtained (table 1), common mechanisms can be utilized as a basis for the electronic decomposition of the γ-HCH pure substance and γ-HCH as a component of the pesticide preparation's active substance. However, as it was noted earlier in our studies, an integrated study is necessary to understand the mechanism of radiation-chemical transformation of substances in a multicomponent system, such as the preparation's suspension. This involves, for example, studying the interaction between active particles formed not only with OCP molecules but also with the inorganic component of both water medium and the filler including natural talc and magnesite.

**Table 1.** Degradation degree of γ-HCCH in samples of the drug dust hexachlorane after electron irradiation.

| D, kGy | P, % | | |
|---|---|---|---|
| | Lindane | Preparation (dust) | Preparation (suspension) |
| 10 | 24.9 | 41.9 | 42.5 |
| 44 | 46.0 | - | 60.0 |
| 100 | 71.0 | 66.3 | 63.5 |
| 120 | - | - | 70.0 |
| 150 | - | - | 85.0 |
| 200 | 71.9 | 68.5 | 67.7 |

During our previous studies [8], it was found that γ-HCH as a pure substance and a component of the pesticide preparation is very much the same in terms of radiation resistance under the exposure of gamma-radiation. The difference determined is of no practical significance. Similar results were obtained for electronic radiation of the substance. That is why, the design of optimal conditions for the destruction of pesticides by using the RT for different radiation sources is possible by the example of complex heterogenous systems, such as the commodity preparation form, which is accumulated in large amounts and is the subject to disposal.

The determination of a differential dependence between the values of P and D helped discover a trend in the destruction activity of γ-HCH in the electronic radiation dose range studied (figure 1).

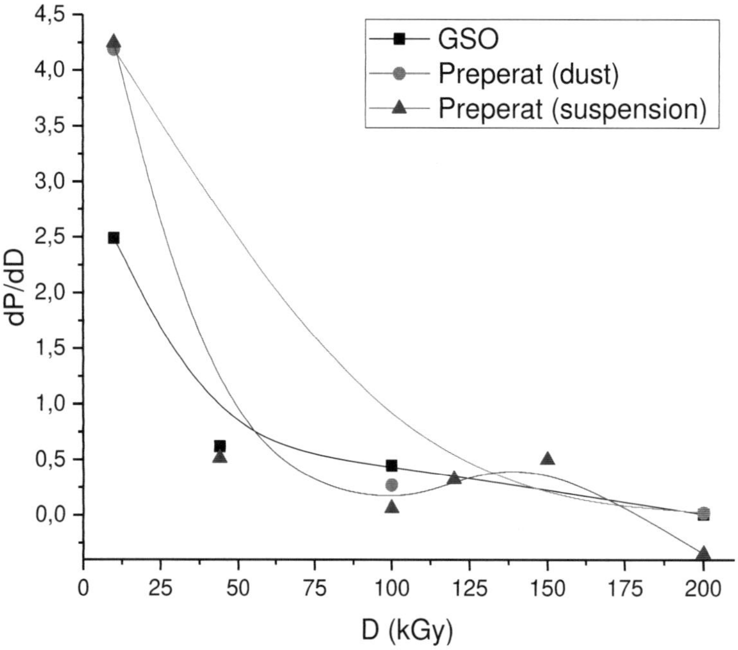


**Figure 1.** Differential dependence curve (ΔP/ΔD(D)) for γ-HCH in the composition of the lindane SRS, the hexachlorane dust (dry powder) and the hexachlorane dust (suspension) irradiated with accelerated electrons

According to the data shown in figure 1, at the impact of accelerated electrons on the samples under study with the doses of 10–200 kGy (electron beam energy of 10 MeV), a degree of decomposition of lindane and γ-HCH as part of the preparation's active substance is the highest at the first stage of radiation exposure until the dose of 10 kGy. When the dose increases to 100 kGy, the activity of the destruction process in the substance slows down and remains unchanged in the range from 100 to 200 kGy. The similar form of the dose dependences for the samples obtained (at different aggregate states) confirms the presence of a common mechanism in a consumption and transformation of the electron beam energy that results in the destruction of chemical bonds in the organochlorine hydrocarbon structure.

We also observed a reduction in the destruction activity of ionising radiation with an increase in the radiation dose when studying the radiation resistance of OCP under the impact of gamma-rays. However, due to the variation of dose levels in the experiments conducted, we did not manage to determine reliably a change in the P values to plot the dose dependence curve.

## 4. Conclusions

It is known that the radical mechanism of radiation-induced destruction of substance is determined by the ratio of rates between two contrary processes: forming active particles under the impact of ionising radiation and developing reactions of their recombination. It can be assumed that in our experiment, we observe a reduction in the rate of the first process. In this connection, the following problems should be solved to intensify the process (for practical purposes):

- the possibility of a rapid increase in the electronic radiation dose for a full mineralization of organochlorine pesticide;
- an application of additional factors enhancing the availability of ionising radiation energy for a decomposed substance on the molecular level, even at small doses of radiation exposure. Such factors can include, for example, reducing the duration of the preparation presence in the active zone, maintaining the medium's turbulence and increasing its temperature and degree of dispersion;
- an expansion of the preparation's aggregate states in the radiation medium; e.g., transferring the preparation's active substance to the vapor phase.

## References

[1] Speranskaya O and Zitzer O Persistent organic pollutants: an overview of the situation in Russia. Available at: https://ipen.org/sites/default/files/documents/4rus_russia_country_situation_report-ru.pdf

[2] Vinokurov A V, Frolov V I and Krestovnikov M P 2010 *Proceedings of RGU Oil and Gas* **2** 61–9

[3] Smirnov V P 2014 Review of the use of radiation technologies *International Forum "Atomic Energy for Sustainable Development"* (NDExpo-2014, St. Petersburg, June 17–19, 2014) Available at: http://dropdoc.ru/doc/154889/obzor-primeneniya-radiacionnyh-tehnologij---ndexpo-2014

[4] Dessouki A M, Abdel Aal S E and Sokker H H 2002 Degradation of some pesticides in aqueous solutions by electron beam and gamma irradiation *7th Conference of Nuclear and Applications*

[5] Pitukhina O A, Polyakova L P, Gordeev A V, Melnikova T V and Oudalova A A 2017 Comparison of the stability of the active substance in "hexachlorane dust" under gamma and electron irradiation *Future of Nuclear Power – AtomFuture 2017: XIII International scientific and practical conference* (Obninsk: NRNU MEPhI) pp 71–3

[6] Lindane. Available at: https://pubchem.ncbi.nlm.nih.gov/compound/lindane#section=Top

[7] 1988 *Hexachlorane dust Technical specifications TU 113-04-254-87* (Ministry for the production of mineral fertilizers) p 13

[8] Polyakova L P, Melnikova T V and Lukyanova N N 2017 *Izvestiya Wysshikh Uchebnykh*

IOP Conf. Series: Materials Science and Engineering **487** (2019) 012024 doi:10.1088/1757-899X/487/1/012024

*Zawedeniy, Yadernaya Energetika* **1** 107–15

[9] Melnikova T, Gordeev A, Polyakova L, Oudalova A, Kozmin G and Naumova L 2017 Alteration of active substance content in "Hexachloran Dust" pesticide formulation at exposure to electron beam *XIII International Youth Scientific and Practical Conference "FUTURE OF ATOMIC ENERGY - AtomFuture 2017, KnE Engineering* pp 433–42

IOP Conf. Series: Materials Science and Engineering **487** (2019) 012025 doi:10.1088/1757-899X/487/1/012025

# Color stabilization in aqueous solution of Beetroot Red

**A Shcherbakova, V Bakhtin, Y Sokolova and P Chelnakova**

Obninsk Institute for Nuclear Power Engineering, National Research Nuclear University MEPhI, Obninsk, Russia

Email: j_sokolova@mail.ru

**Abstract.** Aqueous solutions of Beetroot Red fade fairly quickly, when kept in daylight, due to betanin hydrolysis. Aqueous solution produced out of Beta vulgaris chips is an unstable disperse solution. Protein coagulation results in turbidity and residuum in the solution under storage conditions. Proteins and the colorant form hydrogen bonds. Under coagulation of these bonds, the colorant is released and colorant hydrolysis is initiated. Hindered protein settling leads to the colorant precipitate together with the protein. In order to stabilize the colorant solution, we have used colloid protection that helped the solution retain the color continuously. Guar gum was used as a high-molecular compound to stabilize the color of the solution continuously.

## 1. Introduction

The interest towards natural colorants is fueled by the fact that bright synthetic colorants used in the food industry and pharmaceutics are detrimental to health. There are many colorants produced from natural raw materials, such as curcumin, carmine, anthocyanin colorants, and betalains. Many of them have a non-stable color, so the actual task is to find conditions and stabilizers that will allow them to be used for a long time. Beetroot Red is one of these colorants.

Beetroot Red contains a natural pigment that belongs to the group of betalains [1]. This pigment group has a limited distribution in nature and is well known for coloring the roots of beetroot (Beta vulgaris). After a day in the daylight, an initially clear solution of Beetroot Red begins to grow turbid, resulting in a white residuum, while after five days, the solution becomes discolored. In this form, solutions of Beetroot Red cannot be used. In an aqueous solution, the betanin molecule undergoes hydrolysis [2]. And accordingly, the colorant solution loses its spectral properties.

## 2. Materials and methods

This research involved beetroot chips – large thinly sliced dried strips of peeled beetroot (Beta vulgaris). This form of a natural product was used to reduce the amount of solid inclusions in the solution. The aqueous solution was infused at room temperature and filtered through a fine-pore filter. The concentration of Beetroot Red was identified as 0.007 mol/L. through the use of E-162 calibration chart by ParfaitDecor.

The change in color of the solution was controlled by the change in optical density/turbidity with the help of KFK-3M. The wavelength identified as $\lambda_{max}$ = 540 nm, which corresponds to the maximum absorption of betanin [3].

In order to achieve stabilization, we used edible guar gum, with a concentration of 0.05 g/ml. The guar gum went through preliminary preparation: it was diluted in water with constant stirring to a homogeneous solution. At high concentrations, guar gum is converted to a gel without heating.

A U-shaped vessel was used for electrophoresis. It was filled with a colorant solution, with a current strength of 2 A and a voltage of 2 V.

## 3. Results and discussion

Beet root is a complex multicomponent system, respectively, the solution obtained from the dry product is also a multi-component system. Despite the initially ideal clarity of the solution, it is a disperse system.

To solve the problems of colorant solutions stabilization, the solution was tested for the properties of disperse systems:

1. one of the signature properties of these systems is the Tyndall cone, which was observed in the solution under study;
2. the colorant solution was subjected to electrophoresis, which resulted in solution separation into two parts. That indicates the presence of a molecular charge.

The sediment formed after coagulation was tested with qualitative reactions to proteins, biuret and ninhydrin, both reactions gave a positive result. Therefore, it can be affirmed that it's the proteins that fall into the residuum, when a solution of Beetroot Red is stored at room temperature. Even after thorough filtration, the colorant solution obtained from the beetroot chips, contains fiber particles that provoke coagulation of the proteins, when precipitating. It is noted that the more solids there are in the solution, the faster the colorant degrades in in it. Probably, the protein and betanin molecules form hydrogen bonds. A breach in those bonds results in a loss of stability of the colorant solution, which leads to accelerated hydrolysis along the aldimine bond [4].

To determine the starting point of the hydrolysis, optical density was measured during the first three hours after the solution was prepared. Figure 1 shows that changes in the solution structure appeared immediately after its preparation, which means hydrogen bonds between the protein and the colorant weaken due to formation of protein conglomerates. During this time, the solution itself remains clear. Only when a certain concentration of coagulant is reached, the perceptible coagulation process begins, leading to solution turbidity and visible to the human eye.

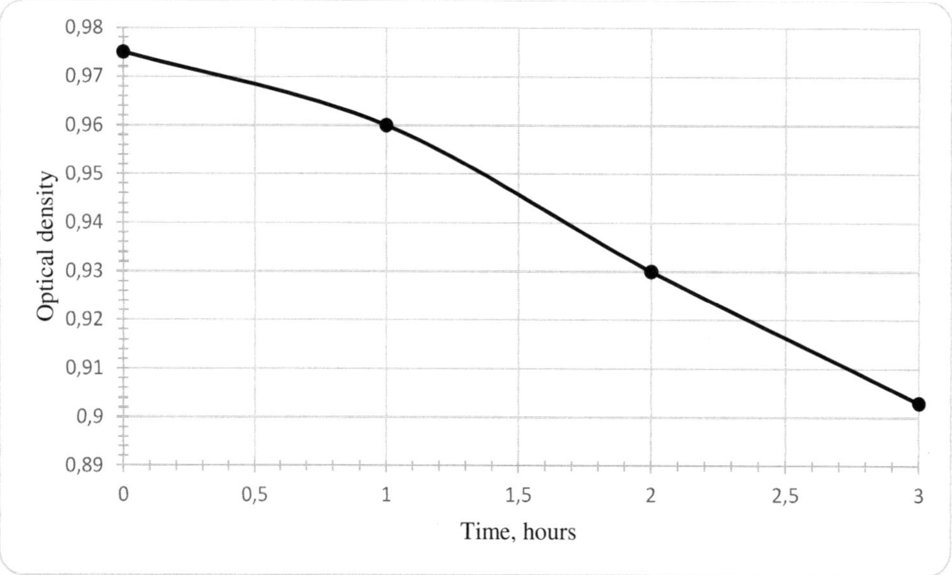


**Figure 1.** Dependence of the initial change in the optical density of Beetroot Red solution on time.

Colloidal protection with a high molecular weight compound (HMWC) was used to stabilize the color of the aqueous solution of Beetroot Red. For this purpose, we used guar gum, a polysaccharide that functions as a thickener in the food industry (figure 2). An increase in the solution viscosity is one of the reasons for its longer storage time.

Its solutions form structured fluids. The gum molecule hydrates through a large number of hydrophilic groups (their presence is indicative for polysaccharides) and stretches into filamentous macromolecules, forming various bonds with protein and colorant molecules, which prevents protein coagulation and, accordingly, betanin decomposition. Since protein and colorant molecules are interconnected, the delay in protein coagulation also leads to an increment in preservation time of the colorant molecule. It should be noted that addition of guar gum results in color preservation, even when a certain amount of protein precipitates.

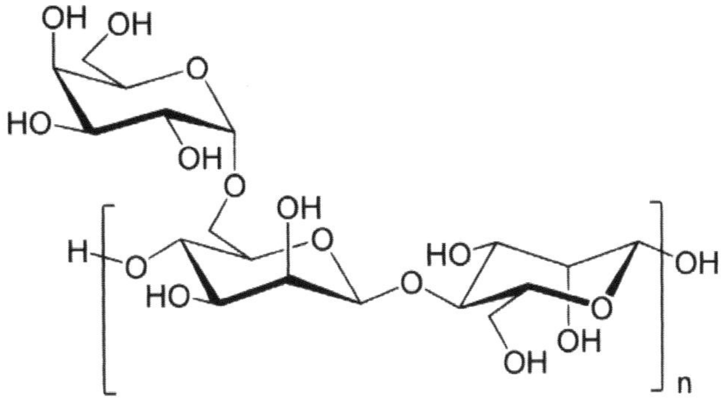


**Figure 2.** Guar gum.

The protein, precipitated about a day after the preparation of the solution, is white. With forced precipitation by lead acetate, the precipitate turns red violet. It is known that a complex of proteins containing lead salts is white, therefore, a colorant is present in the residuum. This confirms bonds between betanin and protein exist. Thus, the protein initiates the colorant hydrolysis, forming large conglomerates, hydrogen bonds between the protein and the colorant are destroyed, which facilitates the attack of water molecules on the massive betanin molecule.

The addition of a dismal amount of guar gum stabilized the colorant solution for a long time. The solution retained its color for three months. In comparison, the usual solution became discolored completely in 120 hours.

Figure 3 shows optical density/turbidity as a function of the solution's exposure to light. Solutions with different guar gum content were used for the experiment. The bend in curve 1 of the colorant-free solution containing no gum is associated with protein coagulation and significant turbidity. It can be seen that after adding 0.1 g/L of guar gum, curve 4 preserves the initial solution's geometry, which indicates a sufficiently strong mutual influence of protein and colorant and the lack of HMWC. At the concentration of 0.2 mol/L (curve 3), the effect is visible, the curve straightens out, i.e. there is no sharp coagulation. The optimum content of guar gum (curve 2) is 1 g/L – the stabilization effect.

To prevent fermentation of the colorant solution, ascorbic acid has been added, which has antioxidant and antibacterial properties. After the acid was added, the following effects were observed: transparency maintained, no fermentation, turbidity or sediment. A different pH of the solution changed the charge of the protein: since the proteins are ampholytes, they precipitate when lose their charge. Therefore, addition of the acid makes the proteins charged and water soluble. But the acidic media adversely affects the gum, since in acid solutions it is hydrolyzed to monosaccharides.

Therefore, figure 4 shows a decrease in the color intensity due to addition of ascorbic acid. But the solution remains sufficiently colored after 250 hours. This experiment showed that a change in pH

IOP Conf. Series: Materials Science and Engineering **487** (2019) 012025 doi:10.1088/1757-899X/487/1/012025

towards acidic values favorably affected stability. Future research will focus on the search for efficient acid additives that do not destroy guar gum and inhibit both coagulation and, accordingly, hydrolysis.

In the colorant solutions with and without guar gum, the charges of the sol were determined by a capillary method through the use of filter paper.

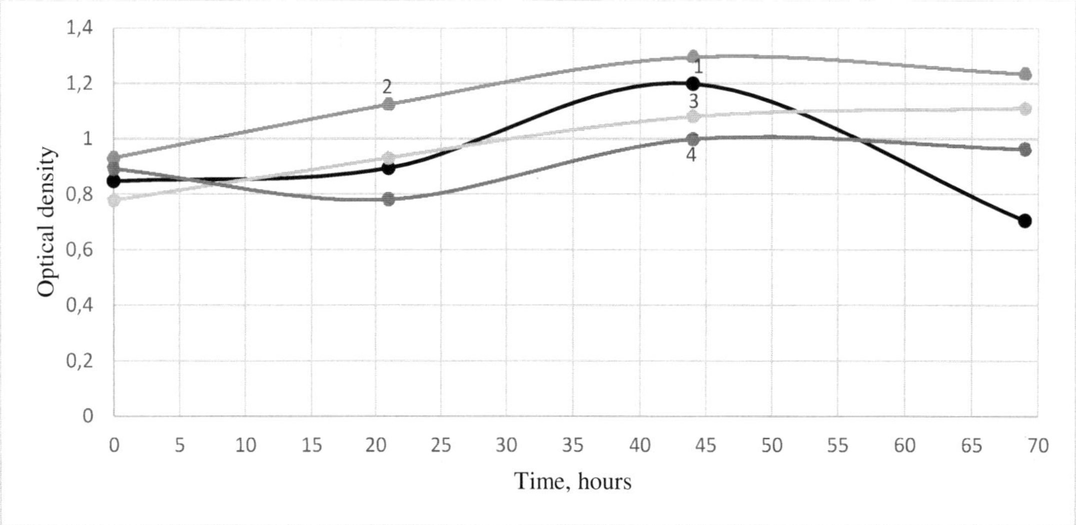


**Figure 3.** Dependence of Beetroot Red aqueous solutions degradation with and without guar gum on time. 1 – concentration of Beetroot Red – 0.007 mol/L; 2 – concentration of Beetroot Red – 0.007 mol/L and concentration of guar gum – 1 g/L; 3 – concentration of Beetroot Red – 0.007 mol/L and concentrations of guar gum – 0.2 g/L; 4 – concentration of Beetroot Red – 0.007 mol/L and concentration of guar gum 0.1 g /L.

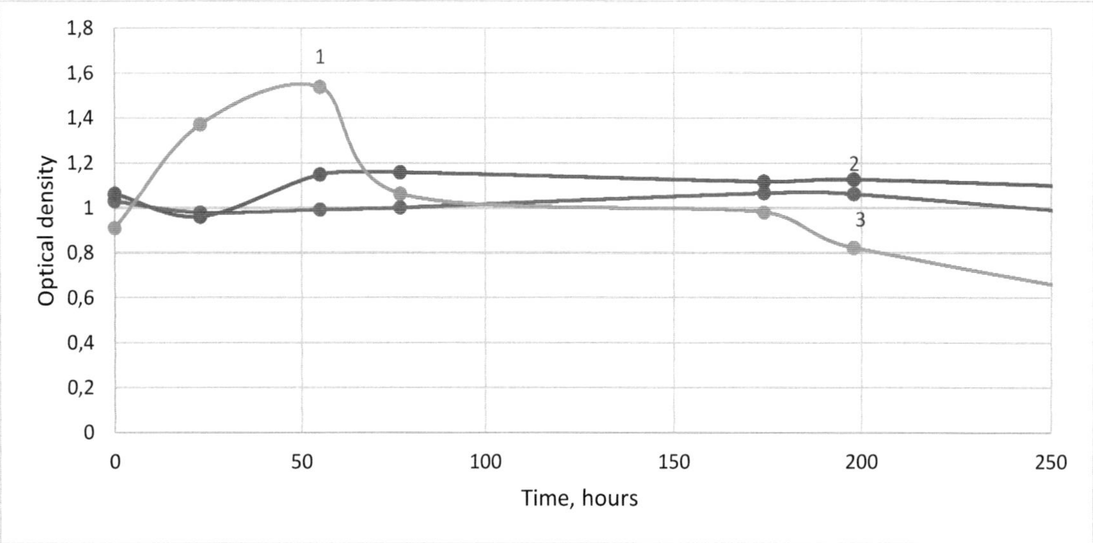


**Figure 4.** Comparison of the degradation time of the Beetroot Red colorant in aqueous solutions with various additives. 1 – concentration: Beetroot Red 0.007 mol/L; 2 – concentration: Beetroot Red 0.007 mol/L, guar gum 1 g/L; 3 – concentration: Beetroot Red 0.007 mol/L, guar gum 1 g/L and ascorbic acid 0.057 mol/L.

Figure 5 shows that both regular solutions and the ones containing guar gum and ascorbic acid have the same charge. Consequently, the additives do not affect the sol charge, and the interaction between guar gum and Beetroot Red is not based on electrostatic attraction.

The following conclusions can made based on this research:

1. It has been established that an aqueous solution of Beetroot Red, obtained from dry beet chips, is an unstable disperse system;
2. Protein coagulation initiates the hydrolysis of the Beetroot Red colorant;
3. Adding small concentrations of guar gum (1 g/L) to the solution stabilizes the color of the solution of the Beetroot Red colorant continuously, due to solution structuring and delay of the hydrolysis;
4. Adding of 1% ascorbic acid allows one to maintain the solution transparency with a satisfactory color retention, as well as to suppress fermentation.

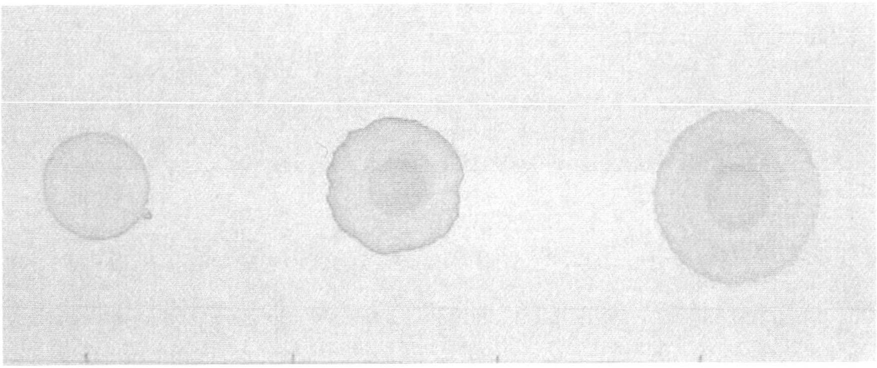

**Figure 5.** Drops of Beetroot Red solutions on filter paper. 1 – solution without additives; 2 – solution containing guar gum; 3 – solution containing a mixture of guar gum and ascorbic acid.

**References**

[1] Britton G 1986 *The Biochemistry of Natural Pigments* (Moscow: Mir) p 442

[2] Sokolova Y 2017 Study of Beetroot Red Fading *Universum: Chemistry and Biology* (Moscow: MTsNMO LLC) **2**(23) pp 47–51

Sokolova Y, Bakhtin V and Chelnakova P 2018 Degradation of Beetroot Red and oxidation reaction in colorant aqueous solutions Proc. the 34th International Research-to-Practice Conference: 'Research and Development 2018' *(Electronic Materials*) (Moscow: Olymp Research Center) pp 40–44

[3] Novikova A, Sokolova Y and Chelnakova P 2016 Study of betalain colorant stability *Proc. of the 13th Regional Research Conf. 'Technogenic systems and environmental risk'* (Obninsk: Obninsk Institute for Nuclear Power Engineering, National Research Nuclear University MEPhI) pp 114–6

[4] Herbach K, Stintzing F C and Carle R 2006 Betalain stability and degradation – structural and chromatic aspects *J Food Sci. (Electronic Materials*) **71** 41–50

# $^{68}$Ga-adsorption on the Si-nanoparticles

**H S Skrabkova[1,3], V B Bubenschikov[2], G E Kodina[2], A S Lunev[2], A A Larenkov[2], N B Epshtein[3] and A V Kabashin[4]**

[1]Chemistry Department, Belarusian State University, Minsk 375-17-226-59-40, Republic of Belarus
[2]Burnasyan Federal Medical Biophysical Center, Moscow, Russia
[3] Obninsk Institute for Nuclear Power Engineering National Research Nuclear University MEPhI, Obninsk, Russia
[4]Aix-Marseille University, Marseille Cedex 9, France

E-mail: skrebkova.a@gmail.com bubenschikov2011@yandex.ru gkodina@yandex.ru

**Abstract**. In recent years, more and more attention has been paid to nanoparticles (NPs) as target drug delivery systems. Silicon-based nanoparticles are one of the most biocompatible nanomaterials that have become widespread. In this research, we studied the possibility of gallium-68 sorption on silicon ($Si\text{-}SiO_x$) and silicon dioxide ($SiO_2$) nanoparticles without their surface modification. Conditions allowing to obtain $^{68}$Ga-NPs with radiochemical yield $\geq$ 90% were found. $^{68}Ga\text{-}SiO_2$ showed much higher stability over $^{68}Ga\text{-}Si\text{-}SiO_x$ in both *in vitro* and *in vivo* experiments.

## 1. Introduction

Nanoparticles (NPs) for biomedical applications have become very popular field of research in the last few decades. Every year the number of papers on this subject increases significantly due to the development of new methods of synthesis of nanomaterials and nanocompositions. Among the large number of nanomaterials, special attention is paid to the NPs of silicon and silicon dioxide. This is due to the fact that they have features such as monodispersity, large surface area, the possibility of high-performance loading of drugs. In addition, their surface can be modified with target ligands or chelating agents [1]. Compared to other nanomaterials, silicon dioxide NPs are biocompatible [2], resistant range to temperature, mechanical stress, and degradation caused by hydrolysis in a wide pH [3]. Thus, silicon dioxide NPs are one of the most frequently used materials in biomedical research [4]. Possibility of their uses, as systems for the delivery of drugs [5], gene transfection [5], contrast agents [6], solutions for radionuclide diagnosis and treatment of tumors [7] has been actively studied in the world. In addition to biomedical applications, silicon dioxide is used in analytical chemistry to separate elements as a stationary phase in chromatographic analysis.

However, one of the problems of $SiO_2$ nanoparticles is the fact that they are not biodegradable and accumulate in various organs, especially in the kidneys and liver, which causes many negative effects. Silicon NPs obtained by laser ablation are Si nanocrystals coated with a thin layer of silicon oxide ($Si\text{-}SiO_x$) [8]. These NPs are biodegradable and considered as a perspective framework for the creation of various drugs [9].

Labeled NPs can be used for lymphoscintigraphy, particularly for intraoperative imaging of sentinel lymph nodes (SLN) [10]. Nanocolloid, labeled with $^{99m}$Tc, is a perfect product for identifying the SLN, however, the search for other radiopharmaceuticals is still in progress all over the world [11].

In addition, available radiopharmaceuticals contain only $^{99m}$Tc, which is used exclusively for the visualization with single photon emission computed tomography. From this perspective the development of other similar radiopharmaceuticals for positron emission tomography (PET) is required.

$^{68}$Ga is one of the most perspective radionuclides for the development of radiopharmaceuticals for PET. It is related to fact that $^{68}$Ga is a generator radionuclide and has optimal nuclear physical characteristics. Gallium-68 decays with the emission of positrons (89%) and by electronic capture (EZ, 11%). The average energy of positrons per decay is 740 keV ($E_{\beta^+ max}$ = 1899 keV) [12]. The half-life is 67.71 minutes. The decay is accompanied by the emission of annihilation γ-quants ($E_{\gamma 1}$ = 511 keV, yield 180 %; $E_{\gamma 2}$ = 1077 keV, yield 2,93 %) [13], figure 1.

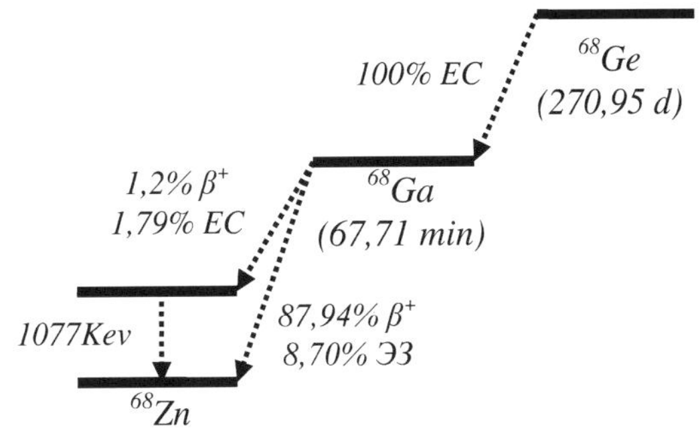


**Figure 1.** $^{68}$Ge decay scheme.

Nowadays, a large number of studies of radioisotope binding to the surface of various NPs, including silicon dioxide, are well known. However, the main part of these studies is based on the modified surface of NPs, whereby the binding of radioisotopes to the surface of NPs occurs through molecular chelating agents [14], whereas the direct adsorption of radioisotopes on NPs is poorly studied [1]. Therefore, the aim of this work was to study the possibility of gallium-68 adsorption on silicon dioxide NPs without modifying the surface.

## 2. Materials and methods

$^{68}$Ga was obtained from $^{68}$Ge/$^{68}$Ga generator ('Cyclotron Co., Ltd.', Obninsk, Russia). The radioactivity was measured with 'Atomlab 500' dose-calibrator (Biodex Medical Systems, Inc., USA). All the experiments were performed using deionized water (18 MOm•cm).

All reagents used in the work belonged to the class of pure ones, particularly: sodium acetate (Macco Organiques, Czech Republic), sodium tartrate (Alfa Aesar, USA), sodium succinate (Sigma-Aldrich), sodium trifluoroacetate (Sigma-Aldrich), sodium phosphate (Panreac, Spain), HEPES (Panreac, Spain), sodium carbonate, sodium hydroxide, hydrochloric acid 0.1 M (Panreac, Spain), ammonium acetate (Panreac, Spain).

Two types of NPs were used: silicon dioxide of 10-20 nm particle size by Sigma-Aldrich ($SiO_2$) and silicon NPs covered with a layer of silicon oxide of 50 nm particle size (Si-$SiO_x$). In experiments on the order of mixing of the reagents two types of $SiO_2$ NPs were used: 10-20 nm sized NPs by Sigma-Aldrich and 100 nm sized NPs 'Polisorb MP'.

Radiolabeling procedure was performed as follows. A buffer solution and suspension containing NPs were placed into an Eppendorf tube followed by adding of the eluate of the $^{68}$Ge/$^{68}$Ga generator. The concentration of NPs in the samples was 5-70 μg/mL. The tubes were incubated in for 15, 30 and

60 minutes at 25, 70 and 95 °C. A control sample (without NP) was prepared and analyzed in parallel with every sample. Radiochemical yield was controlled with radiometry and TLC.

Radiometry was carried out as follows. 10 μL aliquot of every sample was taken immediately after the end of labeling process. Then samples containing NPs were centrifuged for 1-10 minutes at 15k rpm. 10 μL aliquot of supernatant of every sample was taken after centrifugation. The amount of adsorbed $^{68}Ga$ was determined by the formula:

$$A_r = \left(1 - \frac{A_l}{A_g}\right) \cdot 100\% = \frac{A_g - A_l}{A_g} \cdot 100\%$$

where $A_r$ – sediment radioactivity related to the total activity; $A_g$ – of the sample taken before the centrifugation; $A_l$ – radioactivity of the sample taken after the centrifugation.

TLC was performed on ITLC-SG strips (Agilent Technologies, USA). Aqueous solution of citric acid (0.05 M), aqueous solution of sodium citrate (0.1 M, pH 4.6), 4% trifluoroacetic acid (TFA) were used as eluents. In all systems, the free $^{68}Ga$ advances with the solvent front, while the hydrolyzed and associated $^{68}Ga$ NPs remain at the origin: $R_{f\ ^{68}Ga-free} = 0,8 - 1,0$; $R_{f\ ^{68}Ga-associated\ with\ nanoparticles} = 0 - 0,2$; $R_{f\ ^{68}Ga-colloid} = 0 - 0,2$. The radioactivity distribution along the chromatographic strip was evaluated using 'miniGita Star' scanner for thin-layer chromatography ('Raytest', Germany).

*In vivo* experiments were carried out on BALB/C mice. The radiopharmaceutical was administered to laboratory animals intravenously into the tail vein in the amount of 0.1 ml with radioactivity of no more than 20 MBq/ml. The biodegradable was visualized with PET/X-RAY Genisys4 positron emission tomograph for laboratory animals (SofieBioscience, USA). The scanning procedure was performed 60 minutes after intravenous injection. The scanning time was 10 minutes.

## 3. Results

### *3.1. System selection for TLC*

Similarly to the methods described in [1], to estimate the adsorption of $^{68}Ga$ on the surface of NPs we tried different stationary phases (ITLC-SG, Silica gel 60, TLC Cellulose F, Whatman-2) with different eluents (aqueous solutions of citric acid, sodium citrate, sodium acetate, EDTA, DTPA, TFA; mixture of acetonitrile and water, etc.). Three chromatographic systems were chosen: aqueous solution of citric acid, aqueous solution of sodium citrate and 4% (TFA) on ITLC-SG strips. However, when comparing chromatograms of sample with NPs and corresponding control sample, it was noted that the radioactivity at the origin of chromatographic plates for control samples was higher than for analogous samples with silicon dioxide NPs. It can be assumed that when mixing components, gallium is sorbed onto the surface of NPs before the maximum formation of its hydroxyl forms occurs, that is, gallium adsorption takes place in all its forms: $Ga^{3+}$, $Ga(OH)^{2+}$, $Ga(OH)_2^+$, $Ga(OH)_3$, while in control sample gallium is maximally hydrolyzed to form $Ga(OH)_3$. Therefore, it can be assumed that gallium sorbed on silicon dioxide nanoparticles in ionic forms is desorbed by these eluents.

Also, in these systems, a significant discrepancy was found in the evaluation of $^{68}Ga$ adsorption on the surface of silicon dioxide NPs by TLC and radiometry. According to the results of radiometry, radioactivity was concentrated on nanoparticles, while according to TLC the amount of $^{68}Ga$-NPs was significantly lower. The difference in the results between TLC and radiometry can be explained by the fact that the eluting solvent is sufficient to partially desorb gallium from the surface of the nanoparticles. It can be concluded that ionic $^{68}Ga$ is weakly retained on the surface of nanoparticles and transchelated by the eluent. Mostly the results of radiometry were used to estimate the adsorption.

*3.2. Effect of pH on adsorption $^{68}$Ga on NPs*

In the view of the configurational lability of gallium [15], one of the key parameters is the pH of the system. We began to study the adsorption of $^{68}$Ga from experiments in various pH ranges. We observed adsorption in excess of 90% in the pH range from 4.0 to 6.5 in the presence of sodium acetate, as a buffer agent (figure 2). At such pH values, gallium is presented in hydrolyzed forms: $Ga^{3+} \overset{pH\ 3}{\longleftrightarrow} GaOH^{2+} \overset{pH\ 4}{\longleftrightarrow} Ga(OH)_2^+ \overset{pH\ 5.3}{\longleftrightarrow} Ga(OH)_3 \overset{pH\ 8}{\longleftrightarrow} Ga(OH)_4^-$ [16].

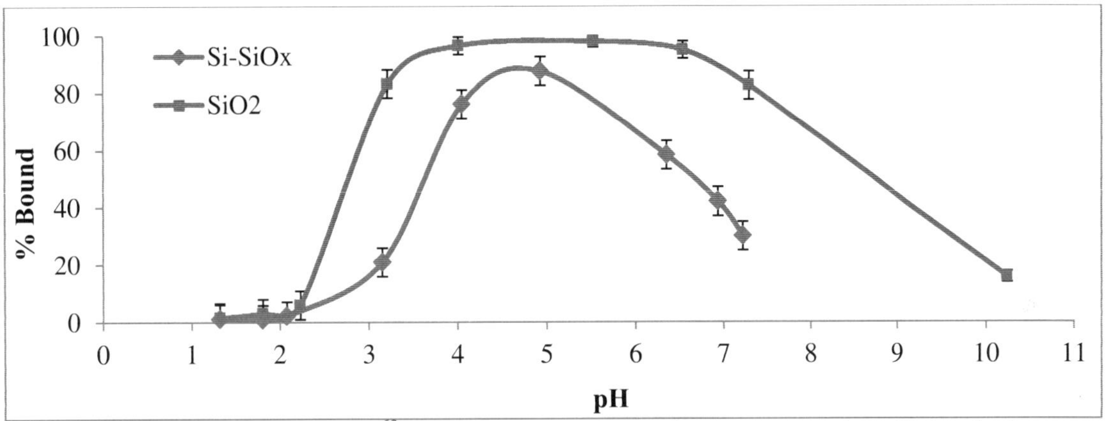


**Figure 2.** Plot of the percentage of $^{68}$Ga bound to particles surface versus reaction pH; incubation 30 min, 70 °C.

The silanol groups on the surface are involved in the adsorption process. Thus, the charge of hydrolyzed forms of gallium, as well as the number of hydroxyl groups, affects gallium adsorption on silicon dioxide nanoparticles. Consequently, it can be assumed that the adsorption is due to the formation of hydrogen bonds of oxygen atoms of hydrolyzed gallium forms with OH groups on the surface of silicon dioxide.

It should be noted that effective adsorption was observed only with the use of sodium acetate, with the maximum adsorption being observed in the pH range of 4.0-6.5 and accompanied by coagulation.

The use of stabilizer (TWEEN 80) positively affected the aggregative stability but led to a drop in adsorption (figure 3).

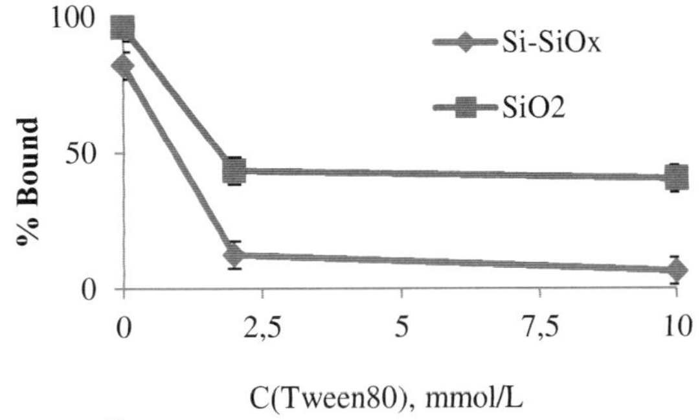


**Figure 3** Plot of the percentage of $^{68}$Ga bound to particles surface versus concentration of TWEEN 80, pH 4.1, 30 minutes, 70 °C.

*3.3. The study of adsorption of $^{68}$Ga colloid forms on NPs*

To select the system in which maximum formation of the $^{68}$Ga-colloid will occur, reaction mixtures in the pH range from 3 to 7 in the presence of various buffer agents (sodium acetate, sodium phosphate, sodium carbonate, sodium succinate, HEPES, ammonium acetate) were considered. The highest yield of $^{68}$Ga-colloid was observed in reaction mixtures with ammonium acetate and sodium carbonate at pH 5. These systems were chosen for further investigation of adsorption of $^{68}$Ga on $SiO_2$-NPs. During the experiments, it was noted that the order of mixing of the reagents affects the adsorption of $^{68}$Ga on nanoparticles. Two orders of mixing of reagents were studied:

I. $^{68}$Ga eluate was added to the NPs solution, followed by a buffer solution. The sample was incubated for 15 minutes at room temperature.

II. $^{68}$Ga eluate was added to the solution of the buffering agent and incubated for 15 minutes at room temperature to form colloid, after which solution of silica nanoparticles was added and again incubated for 15 minutes at room temperature.

The results of the experiment are shown in figure 4.

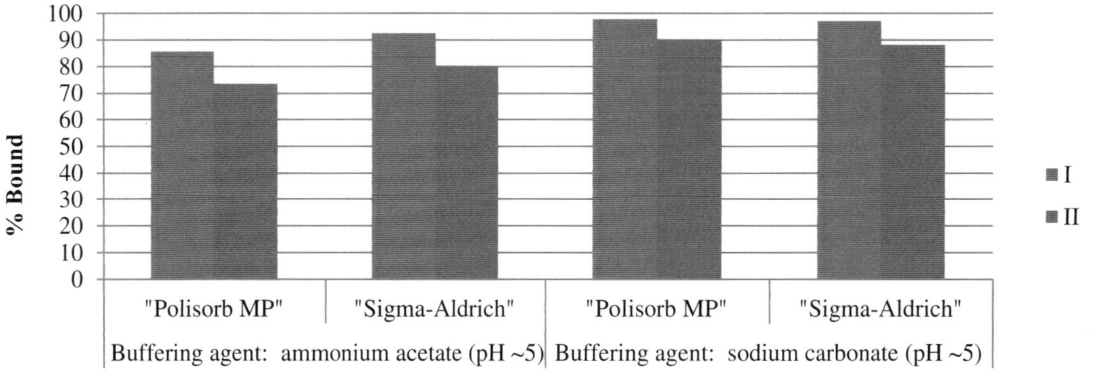


**Figure 4.** Plot of the percentage of $^{68}$Ga bound to particles surface versus the order of mixing of the reagents.

When the order of mixing from I to II is changed, $^{68}$Ga adsorption on nanoparticles decreases, the data is consistent for two types of silicon dioxide nanoparticles, as well as using different buffering agents. It can be assumed that with the order of mixing of reagents, several independent processes occur in the system: the formation of $^{68}$Ga-colloid and its adsorption on the surface of nanoparticles, as well as adsorption of partially hydrolyzed forms of $^{68}$Ga and hydrated ionic forms of $^{68}$Ga. Moreover, in the first order of mixing the reagents on the nanoparticle surface, it is the ionic forms of $^{68}$Ga that are immediately adsorbed, since the nanoparticle solution immediately mixes with the eluate, and only then is a solution of the buffering agent added that affects the formation of the $^{68}$Ga-colloid. From this it can be concluded that the best adsorption of $^{68}$Ga on the surface of nanoparticles occurs during the course of competitive processes, namely adsorption of colloidal and ionic forms of $^{68}$Ga.

*3.4. Comparison of $SiO_2$ NPs with Si-$SiO_x$ NPs*

Further, we carried out comparative experiments on the adsorption of $^{68}$Ga on $SiO_2$, Si-$SiO_x$ nanoparticles. The experiments were performed at pH 7.3, since nanoparticles exhibit high aggregative stability at pH > 7. The results are shown in figure 5–7.

The obtained data allow us to conclude that with an increase in temperature to 95°C there is a decrease in adsorption of $^{68}$Ga on $SiO_2$ nanoparticles.

Despite the rather low $^{68}$Ga adsorption results on Si-$SiO_x$ nanoparticles at pH 7.3, it was decided to carry out a biological study with both $SiO_2$, Si-$SiO_x$ nanoparticles. Nanoparticles labeled with $^{68}$Ga (amount of $SiO_2$ = 0.625 mg/kg of body weight; amount of Si-$SiO_x$ = 0.22 μg/kg of body weight; 2

MBq per mice) were injected to BALB/c mice in the tail vein. 60 minutes after the injection PET images were obtained ('PET/X-RAY Genisys4', Sofie Bioscience, USA). The results of the imaging are presented in figure 8.

Labeled nanoparticles $SiO_2$ accumulate only in the liver and are not excreted in the urine within 60 minutes after the injection, while the free $^{68}Ga$ remains in the blood. This is evidenced by a strong contrast on the images obtained (figure 9a). It should be noted that Si-$SiO_x$ nanoparticles have significant background accumulation in the bloodstream and bladder (figure 9b), which indicates a large amount of free $^{68}Ga$. According to our data, the low thickness of the oxide layer and the lack of a developed surface is the main problem for successful labeling.

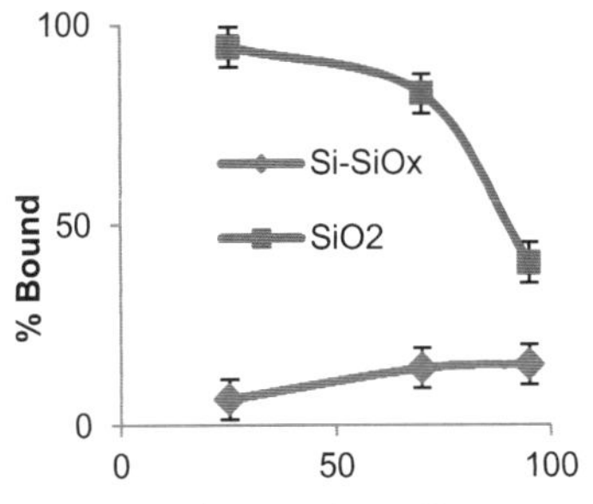


**Figure 5.** Plot of the percentage of $^{68}Ga$ bound to particles surface versus reaction temperature, pH 7.3; 30 min.

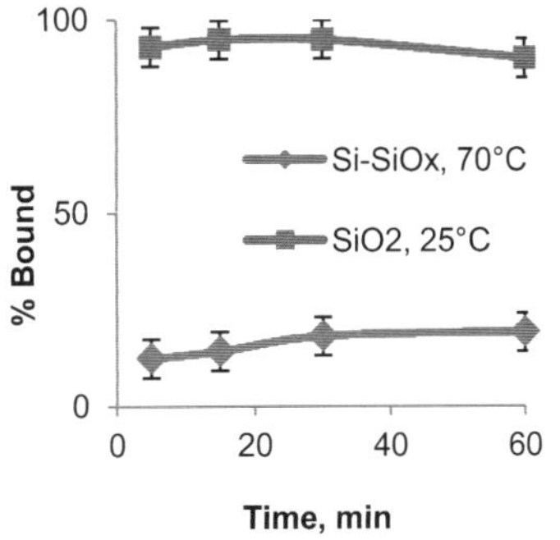


**Figure 6.** Plot of the percentage of $^{68}Ga$ bound to particles surface versus reaction time, pH 7.3; 70 °C.

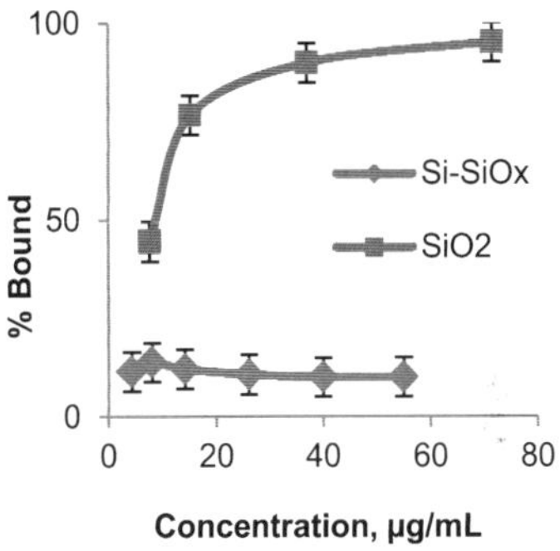


**Figure 7.** Plot of the percentage of $^{68}Ga$ bound to particles surface versus concentration of nanoparticles.

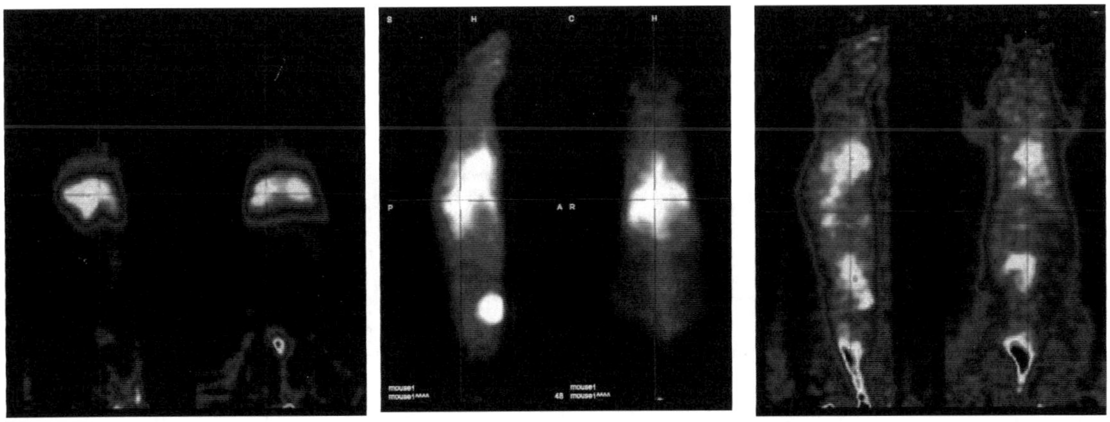

**Figure 8.** In vivo PET imaging (a) $^{68}Ga$-$SiO_2$; (B) $^{68}Ga$-Si-$SiO_x$; (C) free $^{68}Ga$.

## 4. Conclusion

As a result of the study, we conclude that $^{68}Ga$ is better adsorbed on $SiO_2$ nanoparticles than on Si-$SiO_x$. $^{68}Ga$ adsorption on Si-$SiO_x$ nanoparticles occurs in lower quantities and only when accompanied by coagulation, this is due to the low thickness of the oxide layer and the lack of a developed surface in Si-$SiO_x$ nanoparticles. The use of the stabilizer has a positive effect on aggregate stability, but leads to a drop in adsorption. With the increase in temperature to 95°C, $^{68}Ga$ sorption on $SiO_2$ nanoparticles

IOP Conf. Series: Materials Science and Engineering **487** (2019) 012026 doi:10.1088/1757-899X/487/1/012026

decreases, this decline is associated with an increase in the degree of hydrolysis of $^{68}Ga$ and the formation of a large number of colloidal forms. Obtained data and the results of *in vivo* experiments indicate the possibility of sorption of $^{68}Ga$ on the surface of $SiO_2$ and Si-$SiO_x$ nanoparticles without surface modification.

**References**


[1] Shaffer T M, Wall M A, Harmsen S, Longo V A, Drain, C M, Kircher M F and Grimm J 2015 *Nano Lett.* **15** 864–8

[2] Slowing I I, Wu C W, Vivero-Escoto J L and Lin V S 2008 *Adv. Drug. Deliv. Rev.* **60** 1278–88

[3] Kretowski R, Kusaczuk M, Naumowicz M, Kotynska J, Szynaka B and Cechowska-Pasko M 2017 *Nanomaterials (Basel).* **7** 230

[4] Zhou Y, Quan G, Wu Q, Zhang X, Niu B, Wu B, Huang Y, Pan X and Wu C 2018 *Acta Pharm. Sinica. B.* **8** 165–77

[5] Chen M and von Mikecz A 2005 *Exp. Cell. Res.* **305** 51–62

[6] Bradbury M S, Pauliah M, Zanzonico P, Wiesner U and Patel S 2016 *Wiley Interdiscip. Rev. Nanomed. Nanobiotechnol.* **8** 535–53

[7] Barrett T, Choyke P L and Kobayashi H 2006 *Contrast. Media. Mol. Imaging.* **1** 230–45

[8] Gongalsky M B et al. 2016 *Sci. Rep.* **6** 24732

[9] Baati T et al. 2016 *Sci. Rep.* **6** 25400

[10] Vera D R, Wallace A M, Hoh C K and Mattrey R F 2001 *J. Nucl. Med.* **42** 951–9

[11] Hodgson N, Zabel P, Mattar A G, Engel C J, Girvan D and Holliday R 2001 *Ann. Surg. Oncol.* **8** 133–7

[12] Fani M, Andre J P and Maecke H R 2008 *Contrast. Media. Mol. Imaging.* **3** 67–77

[13] Saha Gopal B 2010 *Fyndamentals of Nuclear Pharmacy* (New York: Springer) p 79

[14] Xing Y, Zhao J, Conti P S and Chen K 2014 *Theranostics* **4** 290–306

[15] Taylor M J 1990 *Polyhedron* **9.2-3** 207–14

[16] Wood S A and Samson I M 2006 *Ore Geology Reviews* **28** 57–102

# Diclofenac as a factor in the change of *Rhodococcus* metabolism

**E A Tyumina[1], G A Bazhutin[2], E V Vikhareva[3], A A Selyaninov[4] and I B Ivshina[1,5]**

[1]Perm State National Research University, Perm, Russia
[2]Perm State Humanitarian Pedagogical University, Perm, Russia
[3]Perm State Pharmaceutical Academy, Perm, Russia
[4]Perm State National Research Polytechnic University, Perm, Russia
[5]Institute of Ecology and Genetics of Microorganisms, Ural Branch Russian Academy of Sciences, Perm, Russia

E-mail: elenatyumina@mail.ru

**Abstract**. The non-steroidal anti-inflammatory drug diclofenac (2-[2-(2,6-dichloroanilino)phenyl] acetic acid) is one of the most frequently detected pharmaceuticals in the environment. The ability of actinobacteria of the genus *Rhodococcus* to biodegrade diclofenac (170 µM) as the sodium salt has been revealed. Cells of *R. ruber* IEGM 231 pre-grown with diclofenac (1.7 µM) removed diclofenac (about 50% degradation) within 56 days in the presence of glucose. Diclofenac induced changes in the morphometric characteristics of rhodococci. The products of diclofenac biodegradation did not have pronounced phytotoxicity.

## 1. Introduction

Diclofenac (DCF; $C_{14}H_{11}Cl_2NO_2$; CAS No. 15307-86-5; 2-[2-(2,6-dichloroanilino)phenyl]acetic acid) is one of the most popular non-steroidal anti-inflammatory drugs (NSAIDs) with pronounced analgesic, anti-inflammatory and antipyretic activities. The annual consumption of DCF is 940 tons per year [1]. It is known that in the human body DCF is not completely metabolized, and part of the administered dose is released unchanged [2]. In sewage treatment plants, the efficiency of removal of DCF does not exceed 60% [3]. In this regard, DCF migrates to natural ecosystems and it is the most frequently detected pharmaceutical in the environment [4]. In water and terrestrial ecosystems, DCF occurs in concentrations ranging from 0.02 ng/L to 20.00 µg/L [4]. Data on the detection of DCF in water of Russia are few. In rivers and reservoirs of Moscow DCF was detected at concentrations of 0.025–0.35 ng/L, and in St. Petersburg sewage DCF concentrations were up to 550 ng/L [5, 6].

DCF is an aromatic chlorinated nitrogen-containing compound, a derivative of aromatic carboxylic acid. The molecular structure of DCF causes its toxicity, persistence and high resistance to degradation.

The toxic effects of this NSAID on birds [7], hydrocoles [8], plants [9] have already been observed. In humans, DCF can cause disruptions in liver and gastrointestinal tract, as well as adverse cerebral and cerebellar effects [10, 11].

In connection with the inefficiency of traditional methods of pharmaceutical waste disposal, the search for alternative technologies is relevant. The use of enzymatic activity of microorganisms is promising. However, works devoted to microbial biodegradation of DCF are few. There are a few works on the biodegradation of DCF by *Actinoplanes* [12], *Brevibacterium* [13], *Labrys* [14], and *Raoultella* [3].

Previously, we selected strains of actinobacteria which are active biodestructors of DCF [15]. To understand the mechanisms of DCF biodegradation, the complete genome of DCF degrading bacterium *R. ruber* IEGM 231 which is also active biodegrader of a wide range of complex organic compounds, has been deciphered [16]. Among the revealed abundance (more than 6000) of catabolic genes, genes encoding oxygenases/hydroxylases (75), dioxygenases (45), cytochromes P450 (21), dehydrogenases (285), laccases and other enzymes were found. Using the obtained data, we proceeded from the assumption that the multipurpose oxygenase systems of actinobacteria will catalyze the processes of direct oxidation of DCF, including the introduction of hydroxyl groups in the aromatic ring until the chemical structure of this ecotoxicant is completely destroyed.

In this paper, the results of the study of DCF biodegradation by *R. ruber* IEGM 231 are presented.

## 2. Materials and methods

The strain *R. ruber* IEGM 231 (DDBJ/EMBL/GenBank under accession numbers CCSD01000001 to CCSD01000115) isolated from spring water (Perm, Russia) is an active biodegrader of complex hydrophobic organic compounds, including the analgesic *n*-acetaminophen [15]. Strain is deposited in the Regional Specialized Collection of Alkanotrophic Microorganisms (IEGM, WDCM 768, http://www.iegmcol.ru/strains/rhodoc/ruber/r_ruber231.html).

Experiments on DCF biodegradation were carried out using the mineral medium RS [17]. DCF was added as the sole carbon and energy source at a concentration of 170 μM. As a co-substrate, glucose (28 mM) was added. Rhodococci were grown in LB broth supplemented with 1.7 μM DCF. Broth cultures were centrifuged at 3,000 rpm for 10 min, washed twice with 10 mM potassium phosphate buffer (pH 7.0) and re-suspended in the same buffer to a concentration of $7.5\times10^{8}$ cells/ml.

A mathematical prediction of the duration of DCF biodegradation process was carried out based on data on removal of DCF. As a mathematical model of the change in the DN concentration, a first-order kinetic equation was used (1).

$$dx / dt = -kx \qquad (1)$$

The controls included (a) sterile DCF-containing medium to evaluate abiotic DCF degradation; (b) DCF-containing medium with *Rhodococcus* cells killed by threefold autoclaving at 120 °C for 20 min to evaluate DCF sorption by cell biomass; (c) *Rhodococcus*-inoculated glucose-containing medium without DCF to differentiate bacterial metabolites resulting from DCF and glucose.

The combined atomic-force and confocal laser scanning system including the MFP-3D-BIO™ atomic force microscope (AFM; Asylum Research Inc., USA) and the Olympus Fluo View 1000 confocal laser microscope (Olympus, Japan) was applied.

The DCF concentrations in the cultivation media during the biodegradation process were measured by high-performance liquid chromatography using a LC Prominence chromatograph (Shimadzu, Japan) equipped with a Phenomenex Jupiter® 5u reversed-phase C18 300 A column 250 × 4.6 mm 5 micron (Phenomenex, USA). Mobile phase was phosphate buffer (pH 3.5) – acetonitrile (60:40). Mobile phase flow rate was 1.0 mL/min. DCF was detected at 276 nm.

Products of DCF biodegradation were analyzed by gas chromatography-mass spectrometry (GC-MS) using an Agilent 6890-5973N chromatograph (Agilent Technologies, USA) equipped with a HP-5MS capillary column (30 m, inner diameter, 0.25 mm) operating in the ionization mode with an electronic impact at 70 eV. The extraction of DCF and its metabolites was carried out by chloroform.

The toxicity of DCF and its biodegradation products was assessed by phytotoxicity test [18] using *Avena sativa* L. Biological activity of individual products was predicted by PASS Online.

All experiments were performed in three replicates.

IOP Conf. Series: Materials Science and Engineering **487** (2019) 012027 doi:10.1088/1757-899X/487/1/012027

## 3. Results and discussion

### *3.1. Dynamics and kinetics of DCF biodegradation*

*R. ruber* IEGM 231 was able to use DCF as the sole source of carbon and energy. In order to enhance the DCF bioconversion process, approaches of pre-growing bacteria in the presence of a low concentration of DCF (1.7 μM), as well as fractional supplementation of glucose as an additional source of carbon, were used. It was found that the residual content of DCF in the post-fermentation culture medium of *R. ruber* IEGM 231 on 56 day of the experiment was about 50% (figure 1).

The kinetic curves for the residual concentration of DCF, obtained by least-squares procedure, corresponded to the experimental data (figure 2). The values of the DCF biodegradation rate constant $k$ and the half-life $t_{1/2}$ were $k = 0.012 \pm 0.004$ days$^{-1}$ and $t_{1/2} = 56.82 \pm 1.71$ days. The prediction of the time to decrease DCF concentration to 1% from the initial DCF concentration with a probability of 95% was 526 days, using the hypothesis of the lognormal distribution law for the constant $k$.

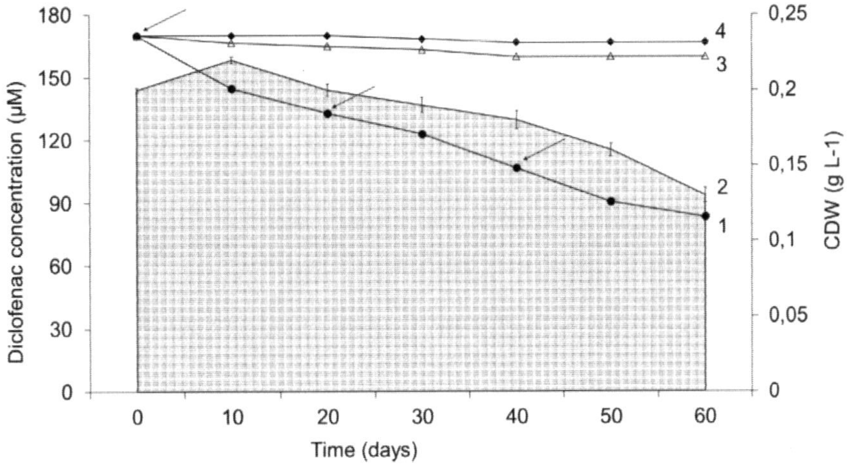


**Figure 1.** Biodegradation of DCF by *R. ruber* IEGM 231 (1). 2 – cellular dry weight, 3 – biosorption control, 4 – abiotic control.

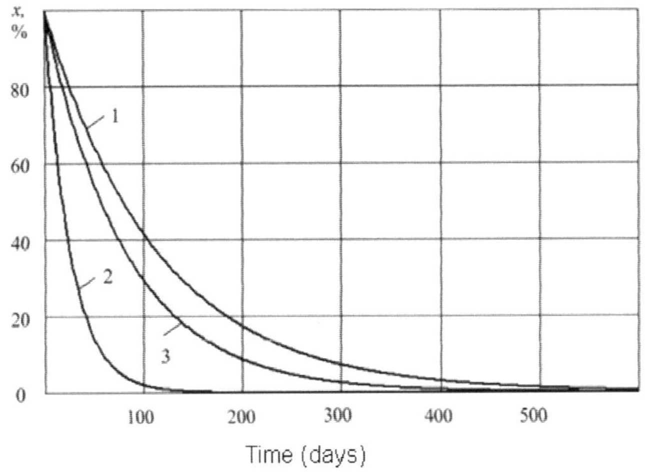


**Figure 2.** Prediction of changes in DCF concentration with a probability of 95% and lognormal distribution law. 1 – upper limit; 2 – lower limit; 3 – selective analog of mathematical expectation.

### *3.2. Change in morphometric characteristics of rhodococci exposed to DCF*

The most typical reactions of *R. ruber* IEGM 231 cells exposed to DCF were an increase in cell length (figure 3B1), a change in the surface profile (figure 3B2), and an increase in the mean square roughness of cell surface (figure 3B3).

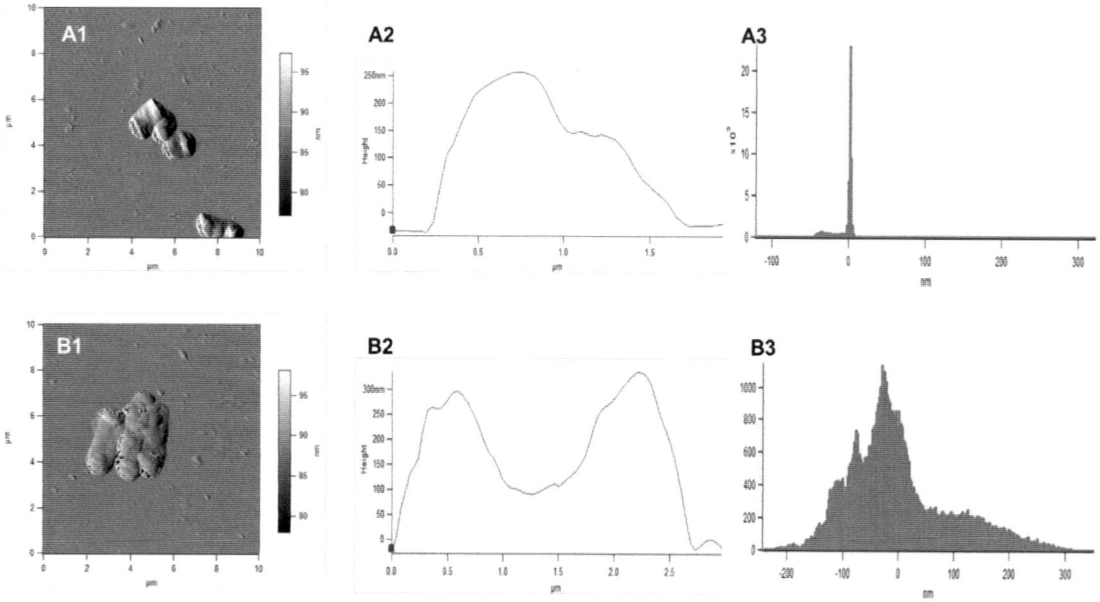


**Figure 3.** AFM-images (1), surface profile (2) and frequency of distribution of surface roughness (3) of *R. ruber* IEGM 231. A – control cells, B – cells exposed to diclofenac.

### *3.3. DCF biodegradation products, their phytotoxicity and biological activity*

The results of GC-MS revealed that among the primary metabolites of DCF biodegradation, 4'-hydroxydiclofenac, 5-hydroxydiclofenac and *p*-benzoquinone imine of 5-hydroxydiclofenac, were found. According to figure 4, they did not show significant phytotoxicity. It should be noted that these compounds undergo further decomposition to simple aliphatic compounds.

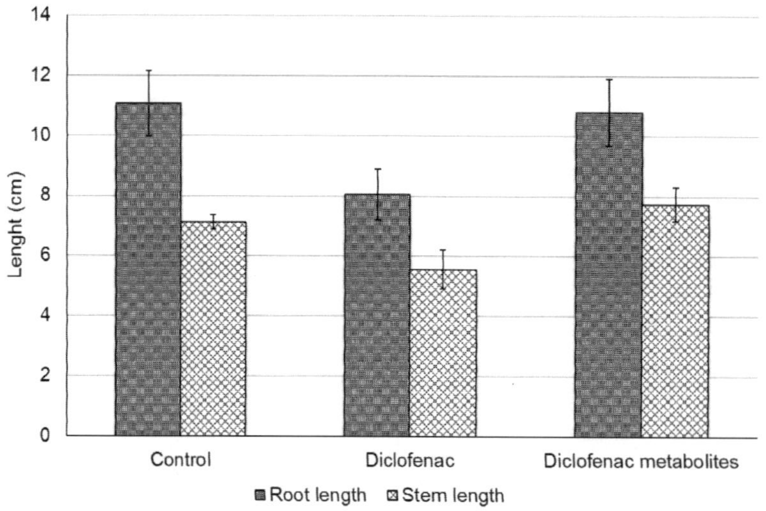


**Figure 4.** Morphometric changes of *Avena sativa* L. exposed to DCF and the products of its biodegradation.

We predicted the biological activity of 5-hydroxydiclofenac of *p*-benzoquinone by PASS Online. Biological activity is described in PASS in a qualitative way ("yes"/"no"). The output of the prediction, in addition to the activity names, includes estimates of the probability of occurrence (Pa) and the absence of each activity (Pi) having values from 0 to 1. Calculations of the prediction of biological activity of *p*-benzoquinone carried out in PASS are summarized in table 1.

It is assumed that *p*-benzoquinone is CYP2J, GST A substrate, has broad inhibitory properties, can be used in the treatment of phobic disorders, rhinitis and can also have antiseboric activity.

**Table 1.** Prediction of biological activity of 5-hydroxiDCF *p*-benzaquinone imine by PASS Online.

| $P_a$ | $P_i$ | **Activity** |
|---|---|---|
| 0,929 | 0,003 | CYP2J2 substrate |
| 0,925 | 0,003 | CYP2J substrate |
| 0,887 | 0,006 | Chlordecone reductase inhibitor |
| 0,837 | 0,009 | Gluconate 2-dehydrogenase (acceptor) inhibitor |
| 0,807 | 0,011 | HIF1A expression inhibitor |
| 0,797 | 0,012 | Anaphylatoxin receptor antagonist |
| 0,758 | 0,021 | Glycosylphosphatidylinositol phospholipase D inhibitor |
| 0,757 | 0,008 | Linoleate diol synthase inhibitor |
| 0,744 | 0,017 | GST A substrate |
| 0,744 | 0,053 | Ubiquinol-cytochrome-c reductase inhibitor |
| 0,728 | 0,023 | NADPH peroxidase inhibitor |
| 0,728 | 0,064 | Phobic disorders treatment |
| 0,719 | 0,003 | Rhinitis treatment |
| 0,707 | 0,036 | Antiseborrheic |
| 0,703 | 0,054 | Membrane integrity agonist |

## 4. Conclusions

We have found found that *R. ruber* IEGM 231 exhibited resistance to a high concentration of DCF (170 μM). Discovered cellular characteristics could be considered as mechanisms of cell adaptation and, as a result, increase their resistance to the negative effects of ecotoxic DCF. During the biodegradation of DCF, products that did not have phytotoxicity were formed.

*R. ruber* IEGM 231 could be recommended for the production of a stable biocatalyst for use in wastewater treatment plants and for bioremediation of contaminated sites from pharmaceutical toxicants.

## Acknowledgments

The study was supported by the Program of Fundamental Research of the Ural Branch of the Russian Academy of Sciences (18-4-8-21) and the Russian Foundation for Basic Research and the Ministry of Education and Science of the Perm Territory (grant number 17-44-590567).

## References

[1] Zhang Y, Geissen S-U and Gal C 2008 *Chemosphere* **73** 1151–61
[2] Vieno N and Sillanpää M 2014 *Environ. Int.* **69** 28–39
[3] Palyzová A, Zahradník J, Marešová H, Sokolová L, Kyslíková E, Grulich M, Štěpánek V, Řezanka T and Kyslík P 2018 *Folia Microbiol. (Praha)* **63** 273–82
[4] aus der Beek T, Weber F A, Bergmann A, Hickmann S, Ebert I, Hein A and Küster A 2016 *Environ. Toxicol. Chem.* **35** 823–35
[5] Barenboim G M, Chiganova M A and Berezovskaya I V 2014 *Water Sector Russia: Probl.*

*Technol. Manag.* **3** 131–41
[6] Russkikh Ya V, Chernova E N, Nikiforov V A and Zhakovskaya Z A 2014 *Ecolog.* **1–2** 77–83
[7] Oaks J L et al 2004 *Nature* **427** 630–33
[8] Guiloski I C, Ribas J L C, da Silva Pereira L, Neves A P P and de Assis H C S 2015 *Ecotoxicol. Environ. Saf.* **114** 204–11
[9] Kummerová M, Zezulka Š, Babula P and Tříska J 2015 *J. Hazard. Mater.* **302** 351–61
[10] Bjorkman D 1998 *Am. J. Med.* **105** 17S–21S
[11] Aygün D, Kaplan S, Odaci E, Onger M E and Altunkaynak M E 2012 *Histol. Histopathol.* **27** 417–36
[12] Osorio-Lozada A, Surapaneni S, Skiles G L and Subramanian R 2008 *Drug Metab. Dispos.* **36** 234–40
[13] Bessa V S, Moreira I S, Tiritan M E and Castro P M L 2017 *Int. Biodeter. Biodegr.* **120** 135–42
[14] Moreira S, Bessa V S, Murgolo S, Piccirillo C, Mascolo G and Castro P M L 2018 *Ecotoxicol. Environ. Saf.* **152** 104–13
[15] Ivshina I B, Tyumina E A and Vikhareva E V 2017 *Microbial biotechnologies: fundamental and applied aspects* vol 9 (Minsk: Belaruskaya Navuka) pp 63–78
[16] Ivshina I B, Kuyukina M S, Krivoruchko A V, Barbe V and Fischer C 2014 *Genome Announc.* **2(6)** e01297–14
[17] Catalogue of Strains of Regional Specialized Collection of Alkanotrophic Microorganisms (accessed May 15rd, 2018. Available from: http://www.iegmcol/strains/index.html)
[18] Liwarska-Bizukojc E and Urbaniak M 2007 *Biotechnologia* **76** 203–14

# Model of the thyroid gland irradiation in the radiobiological experiment analysis

**E N Denisova[1], A S Snegirev[1], V A Budarkov[2], Yu A Kurachenko[1] and G V Kozmin[1]**

[1]Russian Institute of Radiology and Agroecology, Obninsk, Russia
[2]Federal Research Center of Virology and Microbiology, village Volginsky, Russia

E-mail: ellaizaden@mail.ru

**Abstract.** Radiation accidents at nuclear installations are usually accompanied by releases of volatile and biologically dangerous iodine radionuclides into the environment, which can lead to contamination of large areas and cause irradiation of the thyroid gland (TG) in the population and mammals. The reference representatives of biota in radioactive contamination of agricultural ecological systems can be agricultural animals. The basis for the selection of farm animals as reference species is the availability of data of constant veterinary control of animal health indicators and characteristics of their changes due to radiation exposure. To date, the regularities of radiation pathology of the thyroid gland in farm animals with the arrival of radioactive isotopes of iodine have been studied. At the same time, the computational base and software for quantitative analysis of dynamics of formation of radiobiological effects are not sufficiently developed. Using numerical models, namely: a) a compartment model of the $^{131}$I metabolism and (b) precise radiation transport model for the thyroid gland the radiobiological characteristics for calves' thyroid irradiation is derived for the experimental conditions. As a result of experimental and calculated data comparison, critical dose resulting in rapid (within a day) radiation destruction of the parenchymal thyroid tissue was estimated.

## 1. Introduction

Radiation accidents at nuclear installations (nuclear power plants, nuclear chemical plants, research reactors) are usually accompanied by releases of volatile and biologically dangerous iodine radionuclides into the environment, which can lead to contamination of large areas and cause irradiation of the thyroid gland (TG) in the population and mammals.

Along with the objectives of human radiation safety, modern regulatory documents suppose an ecocentric approach to radiation protection [1]. Environmental quality standards in the vast majority of radiological situations are observed if hu-man radiation safety is ensured. At the same time, biological organisms, plants and animals, when polluting the environment with radionuclides, are most often exposed to large doses (comparatively with human doses) the knowledge of which provides additional information for assessing the quality of the environment and making decisions for the preservation of natural ecological systems, the gene pool of plants and animals.

Due to the large biological diversity of living organisms, it is impossible to estimate doses for each type of biota, so it is customary to assess a small set of representative organisms – reference species. These species are considered as biological objects with certain basic eco-logical and biological characteristics. The reference representatives of biota in radioactive contamination of agricultural ecological systems can be rural animals. The basis for the selection of farm animals as reference species is the availability of data of constant veterinary control of animal health indicators and characteristics of their changes due to radiation exposure [2].

To date, the main radiation syndromes have been studied, the manifestation of which in animals can be expected for various scenarios of radioactive contamination. The regularities of radiation pathology of the thyroid gland in the admission of iodine radioactive isotopes were studied. The data on the state of health of animals in the areas of influence of radiation accidents have been received. At the same time, the computational base and modern software for quantitative analysis of the dynamics of the formation of radiobiological effects are not sufficiently developed.

The object of the present study is aimed at estimation the critical radiation exposure leading to radiation destruction of the calves thyroid parenchymal tissue, on the basis of a) compartment model of the 131I metabolism and (b) precise model of the radiation transport in the TG.

## 2. Materials and methods

For estimation of the $^{131}$I effect on the TG of farm animals at radiation accidents the experimental data for cattle's young growth [3] have been used. The calves of 3-month age with weight ~ 60 kg took in $^{131}$I with two liters of milk daily during 6 days: at the first day 185 MBq, the activity of the next portions decreased in accordance with a radionuclide half-life period. On the 11th day of observations there was a sharp decrease in the dose rate of $\gamma$-radiation in the thyroid region [3], possibly due to radiation destruction of the parenchymal tissue of the TG and the $^{131}$I injection into the blood stream. Also, the possibility of radiation-chemical decomposition of thyroxins in the thyroid tissue is not excluded.

Calculation of the $^{131}$I radiation field characteristics was carried out by means of the MCNP5 [4] code. The TG morphology, tissue density, weight and the characteristic sizes are taken from [5].

To estimate the activity of $^{131}$I in TG, a compartment model of the $^{131}$I metabolism in the cattle organisms was used [6], modified in relation to the experimental conditions [3] (figure 1).

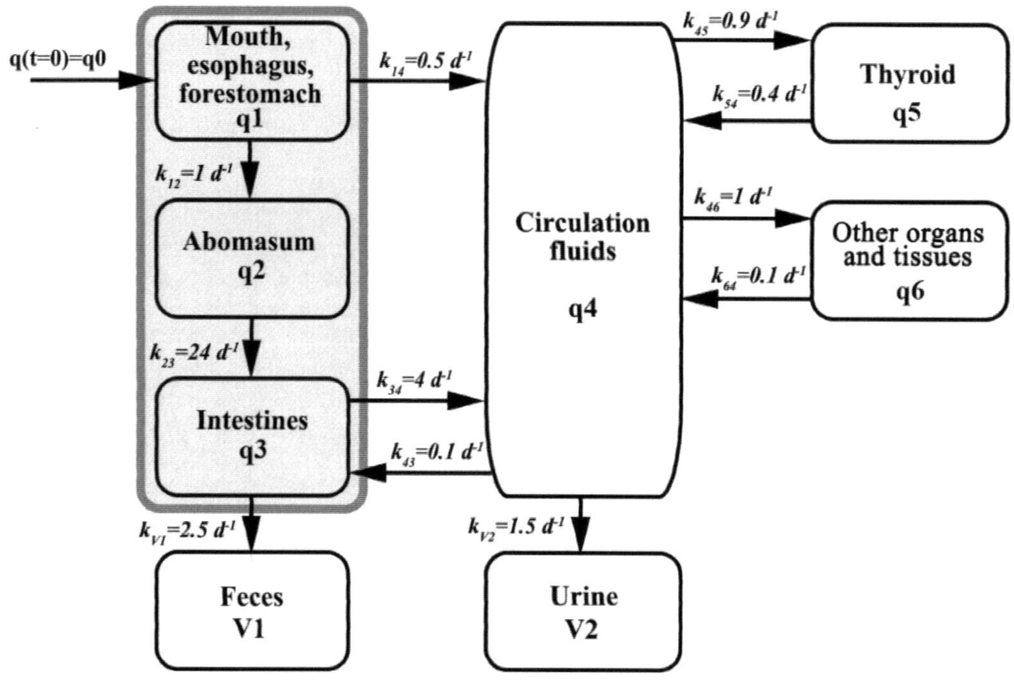


**Figure 1.** Compartment model of the $^{131}$I metabolism in the calves' organism.

In General form, the dynamics of the iodine activity, received peroral, in an arbitrary compartment qi (i = 1...n) under the assumption of identical chemical form of $^{131}I$ in all the compartments during the whole observation time is formulated as follows:

$$\frac{dq_i}{dt} = -\lambda q_i + \sum_j k_{ji} q_j - \sum_j k_{ij} q_i + f(t) \cdot \theta_i$$

where

$q_i$ – the $^{131}I$ amount in the i-th compartment (MBq);
λ – the $^{131}I$ radioactive decay constant (days$^{-1}$);
$k_{ij}$ ($k_{ji}$) – the rate constants of the $^{131}I$ transport for compartment model (days$^{-1}$);

$$\theta_i = \begin{cases} 1, & i = 1 \\ 0, & i > 0 \end{cases}$$

f(t) – rate function of the $^{131}I$ receipt from the external environment in barrier body.

At single oral consumption of the forage containing $^{131}I$, his receipt in barrier body (in this case – the GIT) sets by a starting condition: $q_1$ (t = 0) = $q_0$.

Estimation of a range for rate constants $k_{ij}$ for calves was carried out with use of the parameters received in an experiment on dairy cows [7] without the Udder compartment and also the maximum parameters of metabolism more peculiar to a young animal, according to [8, 9].

## 3. Results and discussion

In figure 2 the TG sections which cover a trachea and consist of the cylindrical layers modeling a form of TG body are presented. According to [3], for calves of 3-month age with weight 60 kg averaged TG volume and weight are 11.6 cm$^3$ and 12.0 g correspondingly; except four main elements, Na, P, S, Cl, K and I enter into element composition. Iodine activity is distributed uniformly in the TG volume.

The current state of computational technologies for the transport of radiation allows accurate modeling of both the subject area and the functionals of the radiation field. The term 'accurate modeling' means that we can put the full extent of our knowledge into the models we construct. This approach in practice leads to a significant narrowing of the error corridor of modeling objects of the subject area and the processes of radiation transport.

| No. | External radius, cm | Height, cm |
|---|---|---|
| 1 | 2.15 | 1.50 |
| 2 | 2.30 | 2.00 |
| 3 | 2.40 | 2.20 |
| 4 | 2.15 | 1.50 |
| 5 | 1.65 | 0.50 |

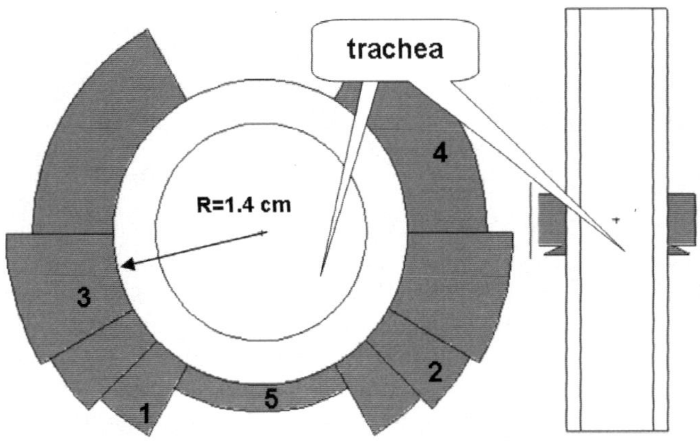


**Figure 2.** Radial (at the left) and axial section of the TG computational model (not to scale; received by visualization of the input file of the MCNP5 code). The table contains dimensions of TG cylindrical layers.

The current state of computational technologies for the transport of radiation allows ac-curate modeling of both the subject area and the functionals of the radiation field. The term "accurate modeling" means that we can put the full extent of our knowledge into the models we construct. This approach in practice leads to a significant narrowing of the error corridor of modeling objects of the subject area and the processes of radiation transport.

The thyroid gland is modeled by a two-dimensional axisymmetric body composed of nine segments, which are obtained by axial and radial sections. These segments are symmetrically located on the trachea and have different outer diameter and height. The same segments are joined in pairs to form four source bodies. The fifth source simulates the isthmus.

The task was formulated as follows:

1. Using the MCNP5 code, designed to calculate the radiation transport by Monte Carlo technique, to obtain an average dose in the model thyroid gland with a uniform volume distribution of the $^{131}I$ activity.

2. Time-varying $^{131}I$ content in the gland was determined using the compartment model.

To calculate total dose in TG, the problems with beta - and gamma-source in each of the five bodies (figure 2) were solved sequentially. Due to the additivity, the dose rates obtained from the two types of source were summed up taking into account mass/volume of the segments. Two components: 1) the dose rate from the beta source and 2) the dose rate from the gamma source combined to give the total dose rate, which was multiplied by the current activity of the gland. As a result, the dependence of the total dose rate on time was obtained. The integral of this dependence over the time interval from the beginning of the experiment allowed calculating the rate of dose accumulation.

Precise calculations were carried out for a) β-radiation transport in TG accompanied by production secondary radiation and its further transport taking into account all processes, including generation and transport of bremsstrahlung, Auger electrons, etc.; b) transport of $^{131}I$ inherent γ-radiation, taking into account generation and transport of x-ray and fluorescent radiations, etc. Energy dissipation was monitored up to 1 keV, integral (energy) dose dispersion was throughout $< 0.1$ %.

It is shown by calculation that in these terms the contribution of γ-radiation produced by $^{131}I$ inherent gamma-source (with all secondary radiations) in the total absorbed dose makes ~ 20%. The main result of calculations was the "conversion coefficient" from $^{131}I$ activity (uniformly distributed in TG) to the average absorbed dose in TG (for these conditions of irradiation): $4.05 \cdot 10^{-12}$ Gy/s per 1 Bq. Accumulative dose in the calves TG during time since which observed destruction of parenchymatous tissue was defined with use of $^{131}I$ activity change in a TG.

Solution of the system of differential equations describing the iodine metabolism according to the presented compartment model (figure 1), by using a multifunctional interactive computing system PTC Mathcad Prime 3.1 has allowed to obtain a change in the $^{131}I$ activity in TG after 6 one-time $^{131}I$ revenues with milk in the gastrointestinal tract of calves. As a result, the dynamics of the absorbed dose accumulation in the thyroid gland of calves was calculated (figure 3).

The value of the absorbed dose accumulated in the TG of calves for the period of time through which rapid (within a day) destruction of parenchymal tissue was observed was ~ 330 Gy.

It is interesting to compare the result with the real situation in the acute period of the radiation accident at the Chernobyl NPP. In lactating cows of Khoyniki and Bragin districts in the Gomel region (the Republic of Belarus) after the Chernobyl accident in May 1986, the concentration of nuclear fission products was 3–4 orders of magnitude higher in the TG of cattle than in other organs and tissues (100–700 kBq/g).

The main factor of radiation danger during the first month after the Chernobyl accident was iodine isotopes. Their content in the mixture of release products ranged from 20 to 60 %. According to gamma-spectrometry, the content of $^{131}I$ in samples of organs and tissues of cows in 5–12 days after the fallout was 65–90 %, and $^{134,137}Cs$ 8–30% of the total activity. The average absorbed doses in the thyroid gland of animals from farms of Bragin and Khoyniki districts was 65.2±22.9 and 164.9±34.6 Gy, respectively [10–12].

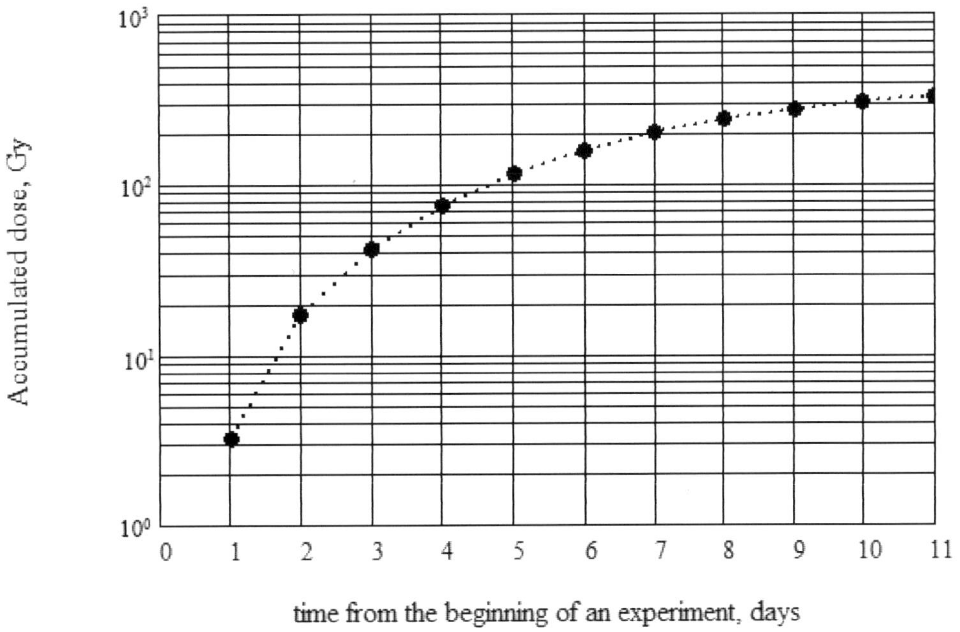


**Figure 3.** The rate of dose accumulation simulated for experiment [3]; after 11 days accumulated dose ~ 330 Gy.

At autopsy of dead animals with maximum thyroid doses, signs of impaired thyroid structure were noted, that for the most polluted farms of Bragin and Khoyniki districts in dynamics manifested in the following way:

- during the first month a sudden impairment of circulation, hyperemia, stromal edema, hypervolemia of the capillaries of inter-follicle tissue, hemorrhage in interstitial tissue and the cavity of the follicle were observed in the TG;
- in the next 1.5–2 months the volume of the thyroid has changed, necronic and dystrophic changes of the follicles' epithelium were founded, pyknosis of nuclei, destruction and desquamation of epithelial cells, the emergence of bands of connective tissue;
- external examination 5 months after the emergency release noted a decrease in the thyroid in volume, its seal and stratification on the cut. Histological examination revealed fibrosis of the organ;
- 8 months later, atrophy of the thyroid was noted. In thyroid' localization the bands of connective tissue appeared.

In the next terms, morphological changes in the TG in the degree of severity were variable in animals of different farms and areas and depended on the density of radioactive contamination of the area and the time elapsed from the beginning of the accident.

The results obtained by us consider a more severe situation due to higher values of $^{131}I$ peroral administered activity and the formation of high levels of thyroid irradiation. Apparently, the lower limit of the dose causing very rapid (within a day) destruction of the parenchymal thyroid tissue of calves corresponds to the value of the absorbed dose obtained in this work ~330 Gy.

## 4. Conclusion

The received results don't contradict the dosimetric data characterizing radiobiological consequences of radiation injury of a thyroid gland (~ 300 Gy) at the person [13] and sheep [11, 12].

The submitted analysis of formation of the absorbed dose in an organism of calves allows estimating the upper bound of the dose causing transient catastrophic destruction of parenchymatous tissue of thyroid gland and the source of $^{131}$I radiation localized in it. The intravital dosimetry of calves at a thyroid glands' region convincingly shows this process: within a day after 11–12 days of observations the dose rate decreases sharply more than in 5 times. By this time in a thyroid gland the absorbed dose is close to 300 Gy.

It should be noted that the destruction of parenchymal thyroid tissue in cattle in the area of influence of the Chernobyl accident was also noted at lower absorbed doses (~150 Gy), but for a longer time (~3–5 months). The mentioned circumstances provide important information for the assessment of animal health in conditions of radioactive contamination of the environment by nuclear fission products.

## References

[1] ICRP Publication 103 2007 The 2007 Recommendations of the International Commission on Radiological Protection *Ann. ICRP* **37** p 332
[2] Budarkov V A 2009 *Radiatsionnaya Biol. Radioekol* **49** 179–85
[3] Spirin Eu V, Lazarev N M and Sarapultcev I A 2004 *Doklady Rossiiskoi Akademii Selskohozyaistvennyh Nauk* **4** 54–5
[4] LANL 2003 *MCNP – A General Monte Carlo N-Particle Transport Code, Version 5*. Vol 1
[5] Suuroja T, Järveots T and Lepp E 2003 *Veterinariya i Zootechnika* **23** 55–9
[6] Snegirev A S, Kozmin G V, Isamov N N, Fesenko S V and Sanzharova N I 2017 *Proc. of research and practice conf. 'Environmental, industrial and energy safety'* (Sevastopol: SevGU) p 1256
[7] Sirotkin A N, Panchenko I J, Tumenev L N, Panteleev L I, Nikolaeva E M, Grishin A I and Korneev N A 1972 *Biologicheskoe deistvie vneshnih i vnutrennih istochnikov radiatsii* (Moscow: Meditcina) pp 72–7
[8] Crout N M J and Voigt G 1996 *J. Dairy Science* **79** 254–9
[9] Smith J G and Simmonds J R 2009 *The Methodology for Assessing the Radiologica Consequences of Routine Releases of Radionuclides to the Environment Used in PC CREAM 08* (Oxfordshire: Radiation Protection Division) p 295
[10] Yastrebkov Yu A, Budarkov V A and Vasilenko I J 1995 *Radiatsionnaya Biol. Radioekol* **35** 845–50
[11] Budarkov V A, Zenkin A C and Arkhipov N I 1992 *Radiobiologiya* **32** 451–8
[12] Budarkov V A, Arkhipov N I and Zenkin A C 1990 *Veterinarija* **7** 60–3
[13] U.S. Dept. HHS 2004 *A Toxicological Profile for Iodine (Atlanta: Agency for Toxic Substances and Disease Registry Division of Toxicology)* p 580

# Activation of the body's natural defenses reserve of mice treated with various physico-chemical agents

**A R Dyukina[1], S I Zaichkina[1], O M Rozanova[1], S S Sorokina[1], S P Romanchenko[1], H N Smirnova[1], A E Shemyakov[1], V A Pikalov[2] and V I Yusupov[3]**

[1]Institute of Theoretical and Experimental Biophysics of RAS, Pushchino, Moscow region, Russia
[2]Institute of High Energy Physics, National Research Center "Kurchatov Institute", Protvino, Moscow region, Russia
[3]Institute of Phonon Technologies, FSRC "Crystallography and Photonics" of RAS, Troitsk, Moscow, Russia

E-mail: dyukina@rambler.ru

**Abstract.** The purpose of this work is to reveal the activation of natural defenses reserve in mice after treatment with different physico-chemical agents *in vivo* using previously developed technology of adaptive response induction. Physical agents were represented by X-rays, carbon ions, infrared light, He-Ne laser light, famine and chemical agents – by immunomodulator $CaCl_2$ and anti-inflammatory drug ibuprofen. The following tasks were set: assessment of cytogenetic damage using a micronucleus test, the weight index of lymphoid organs (thymus and spleen) and the level of ROS production in whole blood through the method of luminol-dependent zymosan-induced chemiluminescence. SHK mice were irradiated according to the scheme of adaptive response. Analysis of data on the number of cytogenetic damage in bone marrow showed that pretreatment of the animals with all investigated agents and subsequent exposure to X-rays or carbon ions at a dose of 1.5 Gy has led to a decrease in radiosensitivity compared to the nontreated animals. Similar results were observed when analyzing weight index of lymphoid organs. Determination of level of ROS production has shown that the activation index calculated according to the relation of induced to spontaneous light area, was significantly higher in all groups of mice, indicating activation of the natural defenses reserve as compared to the group exposed only at a dose of 1.5 Gy. The obtained results confirm the assumption of revealing activation of the natural defense of the organism with the help of the adaptive response induction technology.

## 1. Introduction

The search for effective protective methods from the damaging effects of ionizing radiation is still a pressing task of radiobiology. All known methods of radiation protection are divided according to the time of introduction into radioprotectors, mitigators and therapeutic agents. Radioprotectors have a protective effect when treating animals not more than 30 minutes before or during exposure. Mitigators are introduced immediately after exposure, and the third group concerns the protection with different chemical or physical agents that can reduce radiation damage after 5 or more hours after treatment by activating the body's natural defenses reserve [1]. Scientists are currently intensively researching non-pharmacological treatments for various diseases for the activation of the biological reserve of resistance to various influences that can lead to genetic disorders malignant transformation of cells and emergence of developmental defects in offspring.

We hypothesized that the most promising way for revealing and study this defenses reserve is to use our technology of induction of radiation adaptive response (AR), since radiation damages induced by small exposures are virtually undetectable and can only be detected using large doses of revealing exposure. The AR phenomenon is that prior irradiation of an object in small adjusting doses of low-LET ionizing radiation (γ and X-ray) leads to reduction of sensitivity by 1.5-2 times the subsequent revealing effects of large doses of the same exposure. The data we obtained earlier, mainly with the induction of an AR with the help of radiation exposure, showed that it depends on various conditions: the magnitude and dose rate, the time parameters between the adapting and revealing doses, the radiation quality, and also the individual sensitivity of the organism [2].

With regard to the high-LET ionizing radiation (protons or carbon ions) used in radiotherapy, there is almost no data on their induction of activation of natural protection. We have shown that, in contrast to γ-radiation, preliminary irradiation of mice with secondary radiation from protons with an energy of 70 GeV and π-mesons did not induce a radiation AR [3]. These results supplemented known data obtained in vitro studies on the inability of high- LET ionizing radiation to induce a radiation AR, since it was shown in vitro that secondary radiation induces difficult and unrepaired DNA damage and, to some extent, itself inhibits DNA repair [4]. In addition, the phenomenon when adapting and revealing exposures are factors of various nature is called cross-adaptive response, which can also be seen as one way to detect activation of natural defenses reserve [5, 6].

In recent decades, medical practice has adopted many various instruments based on the use of non-ionizing low-intensity electromagnetic radiation. Infrared light (IRL) and He-Ne laser (HNL) radiation that have bio-stimulating effect are most commonly application [7–12]. Despite these there are few dates on genotoxic effects of these radiations with certain operating parameters, they are controversial and not classified according to power and damage criteria [13–16]. The existing data do not allow concluding about the optimal doses of different radiations for activation of natural protection reserve. Therefore, the problem of finding different agents of physical or chemical nature capable of activation of the body's natural defense, same as small doses of ionizing radiation, is very relevant.

Therefore, the aim of this work is to detect activation of body's natural defenses reserve through radiation and cross adaptations induction technology when treating mice with different physical and chemical adaptogens and revealing doses of low- and high-LET ionizing radiation. The following tasks were set: assessment of cytogenetic damage using a micronucleus test in bone marrow cells, the weight index of lymphoid organs (thymus and spleen) and the level of reactive oxygen species (ROS) production in whole blood.

## 2. Materials and methods

Experiments were performed with male outbred albino SHK mice (body weight 26–30 g) at an age of two months. The animals were kept under the standard conditions in the vivarium of the Institute of Theoretical and Experimental Biophysics (Russia). The experiments conformed to the regulations and legal acts concerning the procedures of animal experiments and the humane treatment of animals.

### *2.1. Experimental condition*

Physical agents for adaption of animals were the following: famine during 24 hours, ionizing radiation – carbon ions with the energy of 450 MeV/nucleon at a dose of 0.1 Gy (uniform pulse beam, 1 times in 8 sec, release duration 0.8 sec / Protvino) and non-ionizing radiation (IRL diode matrix – irradiation of the whole body of the animal for 10 min (850 nm, 22 mW/cm$^2$) and HNL – for 15 and 100 sec (632.8 nm, 0.7 mW, 0.16 mW/cm$^2$), at which the nose of the animal was irradiated.

Chemical agents were represented by immunomodulator $CaCl_2$ (1 g/250 ml of water was given to animals during 6 days) and pharmacological anti-inflammatory drug ibuprofen (i.p. 10 mg/kg once).

One day after the treatments, all groups of animals were additionally irradiated with carbon ions or X-rays at doses of 1.5 Gy according to the earlier used scheme (0.1 Gy + 1.5 Gy) [17].

As a positive control mice were irradiated with X-rays at doses of 0.1 and 0.5 Gy for the radiation AR induction (RUT setup, Mosrentgen, Russia; 0.1 Gy/min, 200 kV, 8 mA, filters 1.0 mm Cu and 1.0 mm Al / Pushchino).

### *2.2. Experimental methods*

Animals were taken from the experiment through decapitation 28 h after exposure to X-rays or carbon ions at a dose of 1.5 Gy, and the bone marrow samples were prepared in order to count polychromatic erythrocytes (PCE) with micronuclei (MN) using standard technique [18]. The experimental point included 5 mice and at least 20000 analyzed PCE. The relative weight of the thymus and spleen were calculated against average absolute organ weight to the average weight of an animal in the group (weight index). Level of ROS production was measured in the whole blood using the method of luminol-dependent zymosan-induced chemiluminescence (ChL) with 12-channel device Chemilum-12 (Russia). Light area was determined under the curve at spontaneous and induced ChL ($S_{sp}$, $S_{ind}$), as well as the activation index (AI), which is the ratio of the $S_{ind}/S_{sp}$ and characterizes the strengthening of ChL. Measurements of ChL in the blood of control and irradiated animals were conducted in parallel. The significance of differences between groups was estimated using the Student's t-test.

## 3. Results and discussion

Table 1 contains data on the number of PCE with MN in the bone marrow of mice treated to IRL, HNL, carbon ions, X-rays, $CaCl_2$ and ibuprofen, with subsequent exposures at a dose of 1.5 Gy of carbon ions or X-rays.

**Table 1.** The number of polychromatic erythrocytes with micronuclei in the bone marrow of mice treated to IRL, HNL, carbon ions, X-rays, $CaCl_2$ and ibuprofen, with subsequent exposures at a dose of 1.5 Gy of carbon ions or X-rays.

| Experimental condition | Number of mice | Number of PCE | Number of PCE with micronuclei | % of PCE with micronuclei |
|---|---|---|---|---|
| 0.1 Gy X-ray + 1.5 Gy C ions | 5 | 20000 | 2768 | 6,9±0,72* |
| 0.1 Gy C ions + 1.5 Gy C ions | 5 | 20000 | 1136 | 5,7±0,36* |
| HNL 15 sec + 1.5 Gy C ions | 5 | 20000 | 1213 | 6,1±0,27* |
| IRL 10 min + 1.5 Gy C ions | 5 | 20000 | 1140 | 5,7±0,42* |
| Famine + 1.5 Gy C ions | 4 | 20000 | 1124 | 5,6±0,35* |
| $CaCl_2$ + 1.5 Gy C ions | 5 | 20000 | 1359 | 6,8±0,46* |
| Ibuprofen + 1.5 Gy C ions | 5 | 20000 | 1297 | 6,5±0,35* |
| 1.5 Gy C ions | 5 | 20000 | 1947 | 9,7±0,65 |
| 0.1 Gy C ions + 1.5 Gy X-ray | 5 | 20000 | 1195 | 5,9±0,41* |
| 0.1 Gy X-ray + 1.5 Gy X-ray | 5 | 20000 | 837 | 4,19±0,72* |
| HNL 15 sec + 1.5 Gy X-ray | 5 | 20000 | 1213 | 6,10±0,45* |
| IRL 10 min + 1.5 Gy X-ray | 5 | 20000 | 1140 | 5,70±0,5* |
| Famine + 1.5 Gy X-ray | 5 | 20000 | 899 | 4,50±0,37* |
| $CaCl_2$ + 1.5 Gy X-ray | 5 | 20000 | 1402 | 6,6±0,47* |
| Ibuprofen + 1.5 Gy X-ray | 5 | 20000 | 1245 | 6,2±0,32* |
| 0.5 Gy X-ray + 1.5 Gy X-ray | 5 | 20000 | 1606 | 8,0±0,42 |
| He-Ne 100 sec + 1.5 Gy X-ray | 5 | 20000 | 1444 | 7,2±0,28 |
| 1.5 Gy X-ray | 5 | 20000 | 1681 | 8,4±0,55 |

* Difference from a group of mice irradiated at a dose of 1.5 Gy, $p < 0.05$

Table 1 shows that the pretreatment of animals with all investigated agents and subsequent exposure to X-rays or carbon ions has led to a decrease in radiosensitivity compared to non-treated animals, which may indicate activation of the body's natural defenses using the cross AR. Preliminary exposure of animal to X-ray at higher doses of 0.5 Gy or HNL radiation during 100 sec and subsequent X-ray exposure at a dose of 1.5 Gy do not lead to a decrease in the number of cytogenetic damage, i.e. activation of the natural protection depends on the dose, as well as the induction of the AR according to the scheme (0.1 Gy + 1.5 Gy). These data are consistent with the results on the induction of AR obtained on mice irradiated with HNL [19].

The weight index of the thymus and spleen were measured simultaneously with the measurement of cytogenetic damage in bone marrow of mice. Thymus and spleen, along with bone marrow, are blood-forming organs with actively proliferating tissue. Figure 1 presents the weight index of thymus and spleen of mice treated to IRL, HNL, carbon ions, X-rays, $CaCl_2$ and ibuprofen, with subsequent exposures at a dose of 1.5 Gy of carbon ions or X-rays. It can be seen that in case of exposure of the untreated mice, thymus and spleen weight index decreases only at a dose of 1.5 Gy. Pretreatment of animal groups with 0.1 Gy of carbon ions, 0.1 Gy of X-ray, HNL 15 sec and IRL10 min has protected thymus from decrease of weight index only when exposed to 1.5 Gy of X-ray, which indicates activation of protective body functions under these conditions. At the same time pretreatment of animals with investigated agents did not restore the weight index of thymus after exposure to these rays compared to the group of untreated animals. Similar results are observed in the analysis of the spleen weight index. Thus, activation of the protection depends not only on the pretreatment quality, but also on the quality of revealing exposure.

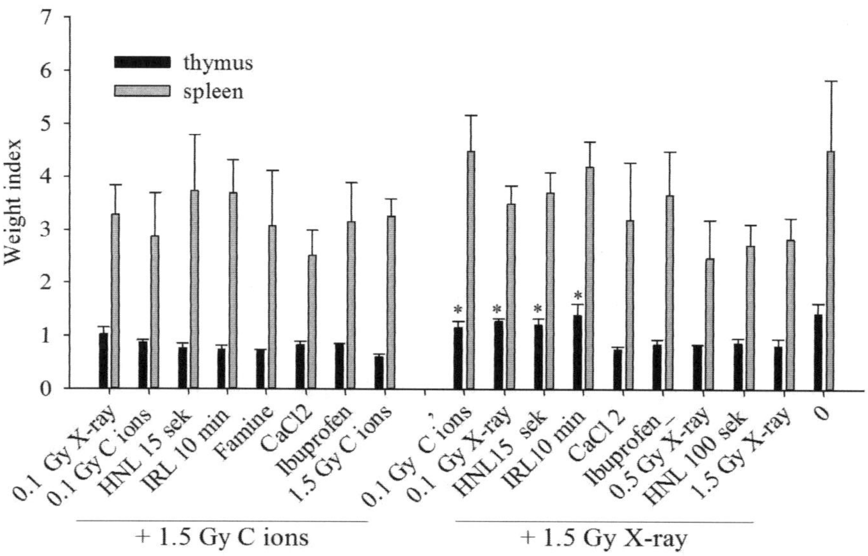


**Figure 1.** The weight index of thymus and spleen of mice treated to IRL, HNL, carbon ions, X-rays, $CaCl_2$ and ibuprofen, with subsequent exposures at a dose of 1.5 Gy of carbon ions or X-rays.
* Difference from a group of mice irradiated at a dose of 1.5 Gy, $p < 0.05$.

It is known that small doses of radiation stimulate weak oxidative stress with the formation of excessive amounts of different ROS. It has been shown in a number of works that the ROS concentration correlates with the activity of the organism [20–22], so the authors examined the level of ROS production in blood cells of the treated and then exposed to 1.5 Gy of X-rays or carbon ions mice. Data for assessing the level of ROS production by neutrophils in the whole blood of mice are presented in figure 2. The amount of spontaneous ChL in neutrophils was measured, which characterizes the basal level of activation of these cells, and the stimulation of oxygen metabolism using opsonized zymosan was performed in order to determine the reserve activation capacity of neutrophils. The activation index was evaluated against the relation of area of ChL induced by zymosan to the area of spontaneous ChL.

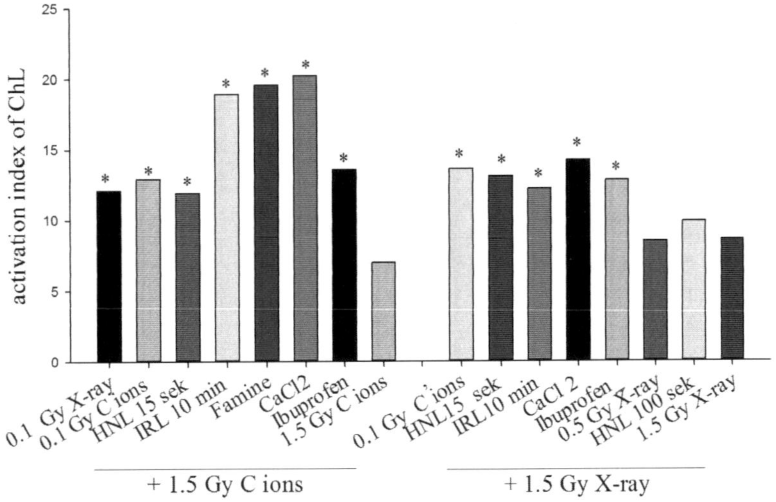


**Figure 2**. The index of chemiluminescence activation of mice treated to IRL, HNL, carbon ions, X-rays, $CaCl_2$ and ibuprofen, with subsequent exposures at a dose of 1.5 Gy of carbon ions or X-rays.
* Difference from the groups of mice irradiated at a dose of 1.5 Gy, $p < 0.05$.

Figure 2 presents the value of chemiluminescence AI for mice treated to all investigated agents, and subsequent exposure at a dose of 1.5 Gy of carbon ions or X-rays. It can be seen from figure 2 that the AI was significantly higher in all groups of pre-treated mice, compared to the group irradiated only at doses of 1.5 Gy of X-rays or carbon ions, indicating the activation of natural protection. At the same time, the preliminary irradiation of animals with X-rays at a dose of 0.5 Gy or HNL radiation for 100 sec and subsequent X-ray exposure at a dose of 1.5 Gy did not increase the AI compared with the group 1.5 Gy, i.e. there was no protection, same as during determination of the level of cytogenetic damage. The effects obtained can be explained by the inclusion of epigenetic mechanisms (signals) that are not associated with a change in the nucleotide sequence of DNA itself and can be indirectly encoded in the genome, which requires further research.

## 4. Conclusion

Thus, the results obtained on identification of methods for protection of mice treated with investigated physico-chemical agents, as well as in the case of induction of the AR, depend on the magnitude of the dose and the quality of the activating and revealing exposures, tissues and methods. These data demonstrate the relationship between primary signals and cellular responses in the hematopoietic and lymphoid organs, and confirm the assumption of activation of the body's natural defense by technology of induction AR.

**Acknowlegements**
This work was partly supported by the Federal Agency of Scientific Organizations (Agreement No. 007-GZ/Ch3363/26).

IOP Conf. Series: Materials Science and Engineering **487** (2019) 012029 doi:10.1088/1757-899X/487/1/012029

**References**

[1] Ilyin L A, Rozhdestvenskij L M, Koterov A N and Borisov N M 2015 *Actual radiobiology: course of lectures* ed. collegium of the series: Smirnov V P et al (M.: Publishing house MPEI) p 240
[2] Zaichkina S I, Rozanova O M, Klokov D Yu, Aptikaeva G F, Akhmadieva A Kh and Smirnova H N 2003 *Radiat. Biol. Radioecology* **43** 153
[3] Klokov D Yu, Zaichkina S I, Rozanova O M, Aptikaeva G F, Akhmadieva A Kh, Smirnova EN and Balakin V Y 2000 *Biological Effects of Low Dose Radiation* eds Yamada T, Mothersill C, Michael B D, Potton C S (Elsevier) p 87.
[4] Antipov A B, Akhmadieva A Kh, Zaichkina S I, Livanova I A, Smirnova E N and Ganassi E E 1993 *Radiat. Biol. Radioecology* **33** 71
[5] Zaichkina S I, Rozanova O M, Aptikaeva G F, Akhmadieva A Kh, Klokov D Yu and Smirnova E N 2001 *Radiat. Biol. Radioecology* **41** 514
[6] Zasukhina G D 2008 *Radiat. Biol. Radioecology* **48** 464
[7] Udagawa Y and Nagasawa H 2000 *In Vivo* **14** 321
[8] Kreisler M, Christoffers A B, Al-Haj H, Willershausen B and d'Hoedt B 2002 *J. Laser. in Surg. and Med.* **30** 365
[9] Toyokawa H, Matsui Y, Uhara J, Tsuchiya H, Teshima S, Nakanishi H, Kwon A H, Azuma Y, Nagaoka T, Ogawa T and Kamiyama Y 2003 *Exp. Biol. Medicine* **228** 724
[10] Mendez T M, Pinheiro A L, Pacheco M T, Nascimento P M and Ramalho L M 2004 *J. Clin Laser Med. Surg.* **22** 19
[11] Bradford A, Barlow A and Chazot L P 2005 *J. Photochem. Photobiol.* **81** 9
[12] Yu S-Y, Chiu J-H, Yang S-D, Hsu Y-C, Lui W-Y and Wu C-W 2006 Photodermatol. Photoimmunol. Photomed **22** 78
[13] Chudnovsky V M, Leonova G N, Skopinov S A and Drozdov A L 2002 *Biological Models and Physical Mechanisms of Laser Therapy* (Vladivostok: Dal'nauka) p 157
[14] Karu T I and Kolyakov S F 2005 *Photomed. Laser Surg.* **23** 355
[15] Karu T I 2007 *Ten Lectures on Basic Science of Laser Phototherapy* (Sweden, Grangesberg: Prima Books AB) p 414
[16] Karu TI 2008 *Photochemistry and Photobiology* **84** 1091
[17] Zaichkina S I, Klokov D Yu, Rozanova O M, Aptikaeva G F and Akhmadieva A Kh 1998 *Russ. J. Genetics* **34** 846
[18] Schmid W 1975 *Mutat. Res.* **31** 9
[19] Zaichkina S I, Dyukina A R, Rozanova O M, Simonova N B, Romanchenko S P, Sorokina S S, Zakrzhevskaya D T, Yusupov V I and Bagratashvili V N 2016 *Bulletin Exp. Biol. Med.* **161** 24
[20] Leach J K, van Tuyle G, Lin P S, Schmidt-Ullrich R and Mikkelsen R B 2001 *Cancer Res.* **61** 3894
[21] Martindale J L and Holbrook N J 2002 *J. Cel. Physiol* **192** 1
[22] David S, O`Shea V and Kundu S 2007 *Nature* **447** 941

IOP Conf. Series: Materials Science and Engineering **487** (2019) 012030 doi:10.1088/1757-899X/487/1/012030

# Assessment of the efficiency of application of ionising radiation for grain disinfestation

**N N Loy, N I Sanzharova and S N Gulina**

Russian Institute of Radiology and Agroecology, Obninsk, Russia

Email: loy.nad@yandex.ru

**Abstract**. The paper considers the results on the efficiency of application of ionising radiation for grain disinfestation. It is shown that gamma-irradiation of confused flour beetle (imago stage) inhabiting on the barley grain, in a dose range 0.7–1.0 kGy caused the 100% mortality of the pest in 15 days after irradiation at a dose rate 0.5 kGy/h. The similar occurred at doses 0.15–0.6 kGy and dose rate 1.8 kGy, and at the dose rate 0.1 kGy/h the total pest mortality occurred in 30 days after the irradiation. The use of the bremsstrahlung and electron radiation for treatment of barley contaminated with confused flour beetle allowed to establish that the mortality of 100% of the insect-pests occurs in 15 days after irradiation at all studied doses and dose rates in case of both radiation types, with exception of the dose 0.25 kGy (bremsstrahlung radiation), when the organisms perish in 30 days after irradiation. The use of gamma-irradiation for radiation disinfestation of grain affects its quality, which depends both on the radiation dose and the dose rate. It was found that the gamma-irradiation caused the increase of the ash content in the grain by 4–6% at doses 0.2–0.6 kGy (dose rate 0.1 kGy/h) and by 9–19.9% at doses 0.75–1.0 kGy (dose rate 1.8 kGy/h). The application of the bremsstrahlung and electron radiation within the investigated dose range had no significant effect on the quality of barley grain of Nur variety. The results obtained on the assessment of the sensitivity of insect-pests (*Tribolium confusum Duv.*) to the effects of different ionising radiation types indicate the prospects of the use of ionising radiation for grain disinfestation.

## 1. Introduction

Insect-pests of grain and grain products are widespread everywhere and cause great harm to agricultural products during storage. They cause not only large losses of grain but also its damage, reduce baking qualities of flour and feeding value of grain fodder and compound feeds.

To ensure secure disinfestation of seeds in Russia they usually use chemicals, particularly contact insecticides. Application of chemical agents (fumigation) is not always effective against the internal contamination of products, and its mass application leads to the emergence of pest forms resistant to the used chemicals agents [1].

In addition, in the process of fumigation, the harmful effects of chemicals agents on the personnel performing treatment are not excluded, and there is practically no possibility of full desorption of the fumigant from the product.

Therefore, during last years the efforts of many scientists and specialists are aimed at finding the new methods of grain decontamination without using pesticides. One of such methods is radiation disinfestation. Use of the gamma-radiation for insect-pests' sterilisation, turned out to be a technically feasible alternative to conventional methods of the pest control in the stored products [2, 3].

Radiation disinfestation of agricultural products in comparison with existing chemical methods has undoubted advantages: environmental pollution is excluded; no pesticides residues in irradiated products [4–6].

Use of radiation treatment is the most promising for the destruction of insect-pests in grain crops. Economic assessments showed that even at a relatively low plant capacity, radiation disinfestation provides greater economic efficiency compared to chemical methods.

The necessary doses for life suppression or control of different pests depend on many factors, and, first of all, on the type of pest and its development stage [7–9].

This work aims to study the efficiency of various doses and types of radiation including gamma, electron and bremsstrahlung, on the viability of the grain pest, confused flour beetle (*Tribolium confusum* Duv.).

## 2. Materials and methods

A subject of study was insect-pest, confused flour beetle (*Tribolium confusum* Duv.) from the *Tenebrionidae* family and spring barley (*Hordeum vulgare* L.) of Nur and Vladimir varieties. Grain, contaminated with pests, was irradiated:

(a) on the facility GUR-120 ($\gamma$-radiation) in the dose range of 0.2–1.0 Gy at dose rates of 0.1, 0.5 and 1.8 kGy/h;

(b) in the bremsstrahlung mode of impulse electron accelerator (bremsstrahlung radiation) in the dose range of 0.25–1.0 kGy at the dose rate 1 Gy/s;

(c) on the electron accelerator (electron radiation) in the dose range 0.5–1.0 kGy (dose rate 1Gy/s).

Modern clinical dosimeter DKS-101 was used to measure absorbed doses in air and objects material. The universal dosimeter DKS-101 is intended for the measurement of absorbed dose and dose equivalent and absorbed dose rate and dose equivalent rate for a wide energy range of photon and electron radiation.

The grain contamination by insect-pests was identified according to GOST 13586.4-83 existing in Russia [10].

The grain quality after treatment (the content of ash, proteins, fat, cellulose, dry matter, nitrogen-free extractive fractions (NFEF)) was determined by the method of diffuse reflective spectroscopy in the near infra-red region of the spectrum on the «Infrapid-61» infra-red analyser. During the analysis, the sample of the refined product was placed in a cuvet with a transparent window and illuminated by radiation with wavelengths lying in the near infra-red region of the spectrum. The analysis time was 2 minutes. It included the readout of the spectrum of the reference standard built into the device, the readout of the spectrum of the analysed sample and processing of the data obtained using an external computer built into the device [11, 12].

The experimental results were processed using Microsoft Excel 2003 software.

## 3. Results and discussions

Radiation dose is one of the leading factors determining the viability of insects at irradiation process. The inhibiting effect of ionising radiation on the functional activity of the organism is directly dependent on the irradiation dose. The value of absolute lethal dose ($LD_{100}$) depends on the insect species and development stage and varies in a wide range [8, 13–15].

Barley grain of Nur variety contaminated with imagos of confused flour beetle was irradiated in a dose range 0.7–1.0 kGy at a dose rate 0.1 kGy/h on the facility GUR-120. The verification of grain contamination in 15 days after irradiation showed that the number of living flour beetle was rather high. The average contamination density was 20–40 organisms per kg (table 1).

During the recording of barley contamination 30 days after the treatment, it was found that flour beetles were completely dead at all studied irradiation doses.

**Table 1.** Influence of gamma-irradiation on barley grain contamination with *Tribolium confusum* (dose rate 0.1 kGy/h).

| Irradiation dose (kGy) | Average contamination density of grain after treatment (organisms per kg) | |
|---|---|---|
| | in 15 days | in 30 days |
| 0 (K) | 20 | 20 |
| 0.70 | 40 | 0 |
| 0.80 | 40 | 0 |
| 0.90 | 20 | 0 |
| 1.00 | 20 | 0 |

Irradiation of barley grain in the same dose range (0.7–1.0 kGy) but at a higher dose rate 0.5 kg/h, was lethal for imago stage of flour beetles even in 15 days after irradiation. This is the evidence of the significance of the dose rate for effect, i.e. dose rate increase leads to a reduction of radiation dose for mortality of 100% of insects (table 2).

**Table 2.** Influence of gamma-irradiation on contamination of barley grain with *Tribolium confusum* (dose rate 0.5 kGy/h).

| Irradiation dose (kGy) | Average contamination density (organisms per kg) | | |
|---|---|---|---|
| | before treatment | after treatment | |
| | | in 15 days | in 30 days |
| 0 (K) | 100 | 100 | 85 |
| 0.70 | 100 | 0 | 0 |
| 0.80 | 60 | 0 | 0 |
| 0.90 | 180 | 0 | 0 |
| 1.00 | 120 | 0 | 0 |

The irradiation efficiency at a dose rate 0.5 kGy/h was 100% even at the first registering of contamination in a period of 15 days after the treatment, while, at a dose rate of 0.1 kGy/h, it occurred only in 30 days after the treatment.

The reduction of the dose range for gamma radiation to 0.15-0.6 kGy and the dose rate increase to 1.8 kGy/h showed 100% efficiency of treatment by a degree of influence on the survival of insects in the contamination determining even in 15 days after the treatment at all studied doses (table 3).

**Table 3.** Influence of gamma-irradiation on contamination of barley grain with confused flour beetle (*Tribolium confusum*) (dose rate 0.5 kGy/h).

| Irradiation dose (kGy) | Average contamination density (organisms per kg) | | |
|---|---|---|---|
| | before treatment | after treatment | |
| | | over 15 days | in 30 days |
| 0 (K) | 200 | 200 | 200 |
| 0.15 | 200 | 0 | 0 |
| 0.20 | 200 | 0 | 0 |
| 0.35 | 200 | 0 | 0 |
| 0.40 | 200 | 0 | 0 |
| 0.60 | 200 | 0 | 0 |

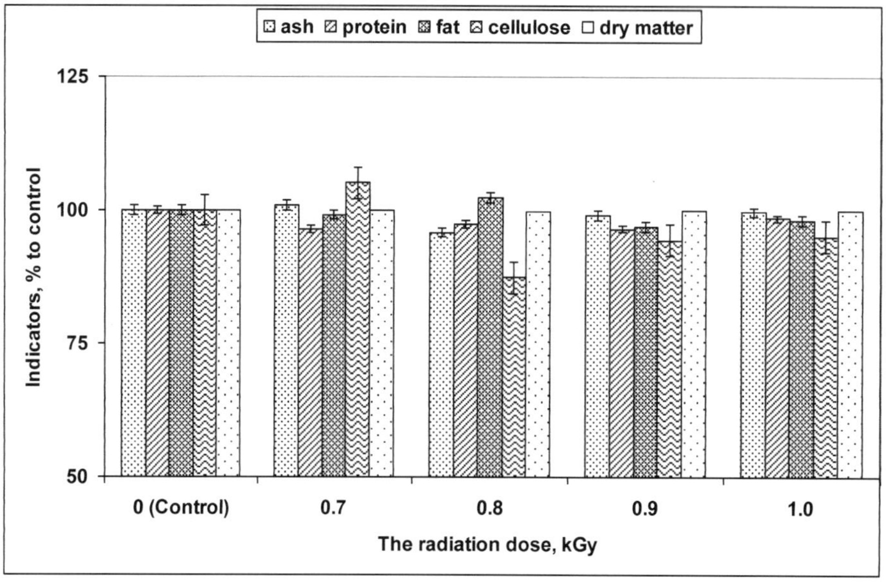


a)

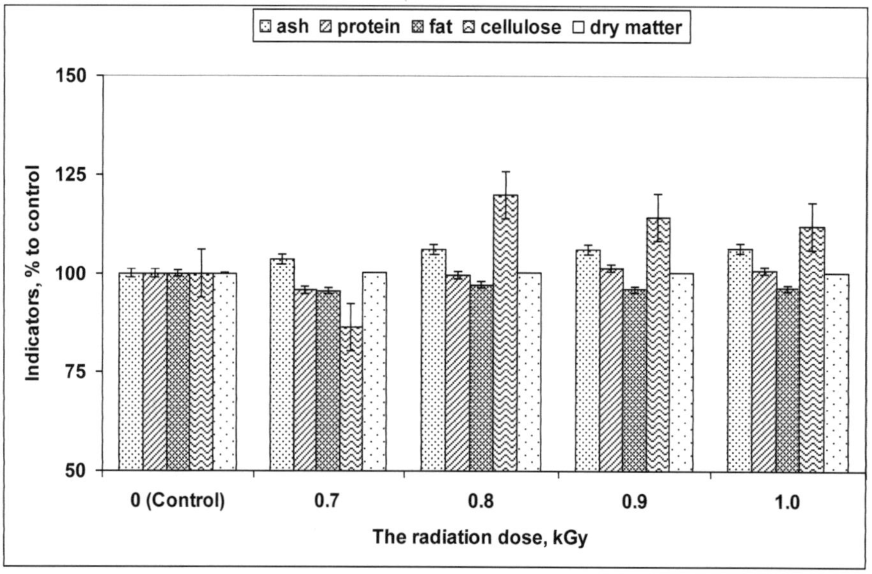


b)

**Figure 1.** Influence of gamma-irradiation on barley quality.
(a) radiation dose rate 0.1 kGy/h, (b) radiation dose rate 0.5 kGy/h).

The application of bremsstrahlung and electron radiation for treatment of barley contaminated with confused flour beetle allowed to establish that the mortality of 100% of the insect-pest occurs in 15 days after irradiation at all studied doses and dose rates and in case of both radiation types, except the dose of 0.25 kGy (bremsstrahlung radiation), when the organisms perish in 30 days after irradiation.

Therefore, the analysis of the grain for the flour beetle contamination after treatment by different types of ionising radiation showed that the insect viability depends on both the radiation dose and the dose rate.

One of the main requirements of ionising radiation application for the treatment of food products is the quality conservation and absence of harmful substances. The studies [16] evaluated the treatment quality of raw materials according to the accepted standards.

Analysis of the grain irradiated to destroy pests at a dose range 0.25–1.0 kGy showed no changes in the content of protein, starch, fat and vitamin complex [17, 18].

Treatment of barley grain of Nur variety in the same dose range but at different dose rate and consequential determination of its quality revealed that at the dose rate 0.1 kGy/h and the radiation dose 0.8 kGy, there was a statistically significant decrease of ash and dry matter content by 4.2 and 0.25%, respectively (figure 1a). While at the dose rate 0.5 kGy/h there was a statistically significant increase of the ash content by 3.6-6.5% at all studied doses and the cellulose content in grain increase by 19.8% at the radiation dose 0.8 kGy (figure 1b).

Barley grain disinfestation in the dose range 0.15–0.5 kGy at the dose rate of 1.8 kGy/h and consequential determination of its quality defined a statistically significant effect of treatment only on the ash content. It increased by 4% at 0.15 and 0.5 kGy doses and decreased by 2.5 and 3.5% at doses 0.3 and 0.4 kGy, respectively (figure 2).

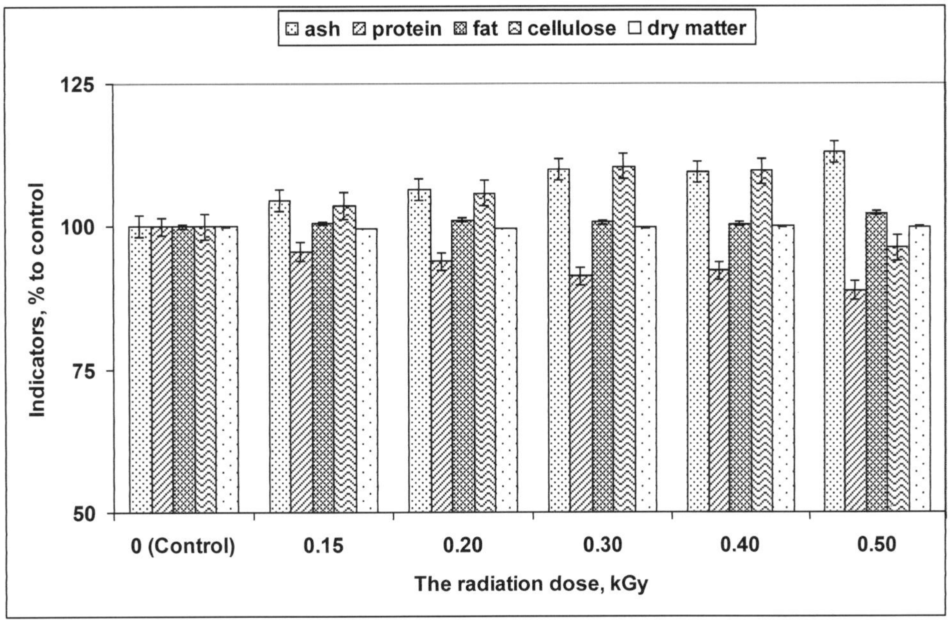


**Figure 2**. Influence of gamma-irradiation on the quality of barley of Nur variety (radiation dose rate 1.8 kGy/h).

Our research found out that varietal characteristics of the crop did not have a significant effect on the irradiated grain quality. Gamma-irradiation of barley grain of Vladimir variety in the dose range 0.17–1.0 kGy and at radiation dose rate 0.1 kGy/h caused a statistically significant increase of the ash content by 6% at doses of 0.17 and 0.3 kGy and its decrease by 5.3% at the dose of 0.9 kGy.

Investigations of the influence of the bremsstrahlung and electron radiation on the quality of barley grain of Nur variety showed that in the studied dose range, both types of radiation did not have a significant effect compared with reference values, except experimental alternative with bremsstrahlung irradiation at the dose of 0.5 kGy. In this case, the content of protein decreased by 5% and fat content decreased by 8%, moreover, at electron irradiation with the dose of 0.75 kGy fat content in grain decreased by 14.6%.

## 4. Conclusions

The research on the effectiveness of the influence of different doses and types of radiation - $\gamma$, electron and bremsstrahlung on the viability of pests from the *Tenebrionidae* family identified that the viability of confused flour beetle depends on both the irradiation dose and dose rate.

Gamma-irradiation of confused flour beetle (imago stage)) inhabiting on the barley grain, in a dose range 0.7–1.0 kGy caused the 100% mortality of the pest in 15 days after irradiation at a dose rate 0.5 kGy/h. The similar occurred at doses 0.15–0.6 kGy and dose rate of 1.8 kGy, and at the dose rate of 0.1 kGy/h the total pest mortality occurred in 30 days after the irradiation.

The use of the bremsstrahlung and electron radiation for treatment of barley contaminated with confused flour beetle allowed to establish that the mortality of 100% of the insect-pests occurs in 15 days after irradiation at all studied doses and dose rates and for both radiation types, except the dose of 0.25 kGy (bremsstrahlung radiation) when the organisms perish in 30 days after irradiation.

Nutrient content in barley grain changed variously at gamma irradiation: at the dose rate 0.1 kGy/h and the radiation dose 0.8 kGy, there was a statistically significant decrease of ash and dry matter content by 4.2 and 0.25%, respectively; at the dose rate 0.5 kGy/h there was a statistically significant increase of the ash content by 3.6–6.5% at all studied doses, and the cellulose content in grain increased by 19.8% at the radiation dose 0.8 kGy.

The chemical composition of irradiated grain does not depend on varietal characteristics of the same crop.

The application of the bremsstrahlung and electron radiation in investigated dose range generally had no significant effect on the quality of barley grain of Nur variety compared with reference values, except experimental alternative with bremsstrahlung irradiation at the dose of 0.5 kGy. In this case, the content of protein decreased by 5% and fat content decreased by 8% (figure 3), moreover, at electron irradiation with the dose of 0.75 kGy fat content in grain decreased by 14.6%.

Therefore, it was established that application of ionising radiation in the dose range of 0.15–0.6 kGy (at the dose rate 0.5 kGy/h and higher) caused a complete suppression of the viability of *Tribolium confusum* Duv. without any reduction of barley grain quality and it can be used for grain and grain products disinfestation.

**References**

[1] Champ B R and Dyte CE 1976 *Report of the FAO Global Survey of Pesticides Susceptibility of Stored Grain Pests* FAO Plant Production and Protection Series vol 5 (Rome: Food and Agriculture Organization of the United Nations) p 297

[2] Sanzharova N I, Geraskin S A, Alexakhin R M, Kozmin G V, Loy N N and Isamov N N (jr) 2013 Prospects of application of radiation technologies in agricultural production *Vestnik Rossiiskoi selskokhozyaistvennoi nauki* **5** 21–4

[3] Loy N N, Sanzharova N I, Gulina S N, Schagina N I, Mironova M P and Gubareva O S 2016 The effect of ionising irradiation on the viability of pests and quality of grain and grain products *Vestnik Rossiiskoi selskokhozyaistvennoi nauki* **6** 53–5

[4] Tuncbilek A S 1995 Effect of $^{60}$Co gamma radiation on the rice weevil, Sitophilus oryzae (L.) *Anzeiger fur Schadlingskunde Pflanzenschutz, Umweltschutz* **68** 37–8

[5] Lapidot M, Saveanu S, Padova R and Ross I 1991 *Insect Disinfestation by Irradiation. Insect Disinfestation of Food and Agricultural Products by Irradiation* (Vienna: IAEA) p 103

[6] Ahmed M 2001 Disinfestation of stored grain, pulses, dried fruits and nuts, and other dried foods *Food Irradiation Principles and Applications* ed R Molins (New York: Wiley) pp 77–112

[7] Brown G A, Brower J H and Tilton E W 1972 Gamma radiation effects on Sitophilus zeamais and S. granarius *J. Economic Entomology* **65** 203–5

[8] Tilton E W and Brower J H 1987 Ionizing radiation for insect control in grain and grain products *Cereal Food World* **32** 330–5

[9] Zewar M M and Abdel-Salam C A 1988 Food consumption by irradiated Sitophilus granarius (L.) and Rhyzopertha dominica F *Agricultural Research Review* **66** 55–61

[10] 1983 *Grain. Methods of Determination of Pest Contamination and Damage. GOST 13586.4-83* (M.: Standardinform)

[11] 2001 *Practical Course on Agrochemistry: Educational Guidance* ed V G Mineyev (M.:

Publishing office of MSU) 24–8
[12] Sazonov Yu G, Pankratov K G and Shchelokov V I 1999 *The Spectroscopy Method in the Near Infrared Region for the Analysis of Agricultural Objects* (M.: Central Institute of Agrochemical Service to Agriculture)
[13] Aldryhim Y N and Adam E E 1999 Efficacy of gamma irradiation against Sitophilus granaries (L.) (Coleoptera: Curculionidae) *J. Stored Products Research* **35** 225–32
[14] Hosseinzadeh A, Shayesteh N, Zolfagharieh H R, Babaei M, Zareshahi H, Mostafavi H A and Fatollahi H 2010 Gamma radiation sensitivity of different stages of saw-toothed grain beetle Oryzaephilus surinamensis L. (Coleoptera: Silvanidae) *J. Plant Prot. Res.* **50**
[15] Hallman G J 2013 Control of stored product pests by ionizing radiation *J. Stored Products Res.* **52** 36–41
[16] *Unified Sanitary Epidemiological and Hygienic Requirements for Goods Subject to Sanitary and Epidemiological Control* Approved by the decision of Customs Union Commission of 28.05.2010. No. 299. Chapter 2. Section 1. Safety requirements and nutritional value of food products pp 6–357
[17] 1995 *Safety and Nutritional Value of Irradiated Products* (M.: Medicine) p 209
[18] Reports published by Department of Atomic Energy. https://docviewer.yandex.ru/?url=https%3A%2F%2Fwww.iaea.org%2FAbout%2FPolicy%2FG.

# Experimental and theoretical assessment of power frequency electric field individual protective means

**S Yu Perov, O V Belaya and T A Khrennikova**

FSBSI «Izmerov Research Institute of Occupational Health», Moscow, Russia

E-mail: sergey.perov@gmail.com

**Abstract**. Electromagnetic field levels at workplaces may exceed permissible limit values. In this case the individual protective means use obligatory. There are presented the date of typical samples of conductive suits (under work in power frequency electric field) testing by simulation and in experiment. Individual protective means screening factors were changed from 42.26 to 52.43 dB. The averaged electric field level inside of conductive suit was 220 mV/m in case of simulation. The induced electric fields in human body tissues with conductive suit were 0.003-0.150 V/m. The results show that conductive suit use allows reducing the induced electric filed levels and ensures worker safety.

## 1. Introduction

One of the actual issues of occupational and environmental safety is human health protection from the adverse influence of various factors, electromagnetic field including ensuring. Electromagnetic field (EMF) levels at workplaces may exceed permissible limit values (PLV) – hygienic standards compliance with which ensures the staff health preservation. In case of EMF levels PLV elevated additional approaches and methods of personnel protection are required. Occasionally the only possible way to protect worker from EMF exposure is individual protective means (IPM) use.

Occupational health and safety under maintenance and exploitation of electrical installations require the elimination of potentially adverse human health effects of power frequency EMF exposure. The protection from electric field (EF) under maintenance and exploitation of electrical installations is provided by IPM use primarily. An important task to ensure safe occupational conditions near the elements of power transmission systems (extremely and ultra high voltage overhead transmission lines, open and closed switchgears) is to reduce the high intensity power frequency (PF) EF (50/60 Hz) lower than PLV by IPM use.

The basis for occupational safety ensuring is PF EF levels compliance to PLV according to hygienic safety standards. According hygienic criteria as protection from for human health unfavorable factor, IPM should provide the required EF intensity attenuation. At the present time, there are quantitative and qualitative differences in the approaches to EF exposure assessment and regulation, which are established by national and international safety standards.

In accordance with the SanPiN 2.2.4.3359-16 «Sanitary-epidemiological requirements for physical factors in the workplace» [1] PF EF permissible limit values in the Russian Federation are depended on time of occupational exposure. Russian hygienic standards of 50 Hz EF are among the most "strict" of the entire world, depending on exposure time. PLV are from 5 kV/m to 25 kV/m (from all work day up to 10 minutes per shift) as is shown in table 1. In case of EF level is higher 25 kV/m the use of protective equipment is obligatory [2].

Individual protective means include clothes (overalls or jacket, trousers/semi overalls), head protection, gloves and/or mittens and boots, made of electrically conductive materials. At the same time, all IPM elements must be electrically connected to each other, forming a closed shell around human body, EF penetration preventing, according to the principle of "Faraday cells".

According to international approaches PF EF are regulated not only by incident EF value but induced electric field strength too.

**Table 1.** Power frequency (50 Hz) electric field occupational exposure permissible limits values according to International and Russian hygienic requirements.

| Exposure value | SanPiN2.2.4.3359-16 [1] | ICNIRP 2010 [3] | Directive 2013/35/EU [4] |
|---|---|---|---|
| Incident EF (kV/m) | 5 (all work day)<br>5-20 (depending on time)<br>25 (≤10 min) | 10 | 10<br>20 |
| Internal electric field (mV/m):<br>Low value<br>(CNS tissue of the head)<br>High value<br>(All tissue of the body and head) | - | 100<br>800 | 140<br>1100 |

The main IPM (conductive suit) characteristic is the shielding coefficient ($K_s$). $K_s$ is attenuation external EF (possible hazard) under IPM to safety levels [5, 6]. EF IPM properties must comply with TR TS 019/2011 «On the safety of personal protective means» [5] and GOST 12.4.172-2014 «Set of individual shielding for protection against electric fields of power frequency. General technical requirements and control methods» requirements, defining 30 dB as minimum permissible shielding factor [6].

The goal of this study was to assess the protective properties of typical samples of modern IPM (EP-1 and EP-3 types), designed for staff protection from PF EF, which takes into account the requirements of national and international safety regulations and guidelines. The objectives of the study included laboratory measurements of IPM shielding properties and the IPM simulation of laboratory test conditions.

In this regard, to assess the IPM protective properties evaluated the EF levels inside and outside of IPM, as well as EF magnitude induced in the human body.

The shielding properties of 4 IPM samples for protection of personnel from PF EF (2 EP-1 and 2 EP-3) were evaluated for work on the ground (ground potential). Laboratory testing of IPM protective properties, the possibility of EF attenuation by the shielding suit were evaluated. Then, using simulation the levels of induced EF in the human body IPM shielding coefficient was obtained as a result of tests.

## 2. Materials and methods

Tests of the IPM shielding properties were carried out on the high-voltage EF power frequency test setup. The setup is open air condenser, with plate sizes of 2.4x1.8 $m^2$, the distance between the plates is 0.7 m. The high-voltage setup provides a uniform electric field in the area of the placement of the human body phantom dressed in the tested IPM. EF exposure levels were from 5 to 85 kV/m. The EF measurement was carried out by a portable electric field analyzer EFA-300 (Narda, USA) with calibrated isotropic electric probe using a fiber optic cable. The probe was placed in a free space inside the body human body phantom. EF values were determined under levels of the external EF, simulated the possible occupational exposure. EF measurements were carried out on a body phantom without IPM, and then on phantom dressed in conductive suit. The screening factor was estimated according to the formula (1), where $E_1$ is the results of electric field measurements without protective suit and $E_2$ - with a suit:

$$K_s = 20\lg\frac{E_1}{E_2} \tag{1}$$

Simulations were carried out using the software product SEMCAD X v.14.8 («SPEAG AG», Switzerland). The Finite-Difference Time-Domain method (FDTD) was used for the calculation for the low frequency range. The resolution of the numerical models was constant for two type of IPM (EP-1 and EP-3).

In numerical simulation to experimental test conditions reproduce, the models of the setup, including a plane-parallel capacitor, screening set and human body phantom, were developed. The capacitor consists of two metal plates, the size corresponding to the real test setup. The upper plate was set to provide a sufficiently uniform EF with 70 kV/m voltage in the work area of the test phantom. The bottom plate was grounded. As real body phantom, a numerically heterogeneous phantom of the human body was used, with 1.77 m high and 72.4 kg weight from Virtual Family [7]. In the simulation used in the conventional dosimetric modeling of the dielectric properties of the tissues of the human body (electrical conductivity and dielectric permittivity) in the tested frequency range [7]. The model of the shielding set is represented by a set of cylinders of different diameters from 2 to 4 mm thick, connected to each other and roughly repeated the shape of the human body. The material of the elements of the set was set as a metal with zero potential.

## 3. Results and discussion

After laboratory tests were obtained averaged values of Ks for IPM different samples (figure 1). Under PF EF levels from 5 to 85 kV/m average Ks was 48.73 dB (EP-1), 52.43 dB (EP-3) and 42.27 dB, 51.25 dB (washed EP-1 and EP-3). Noticeably decrease Ks was mentioned after IPM washing. EP-3 shielding properties was better than EP-1.

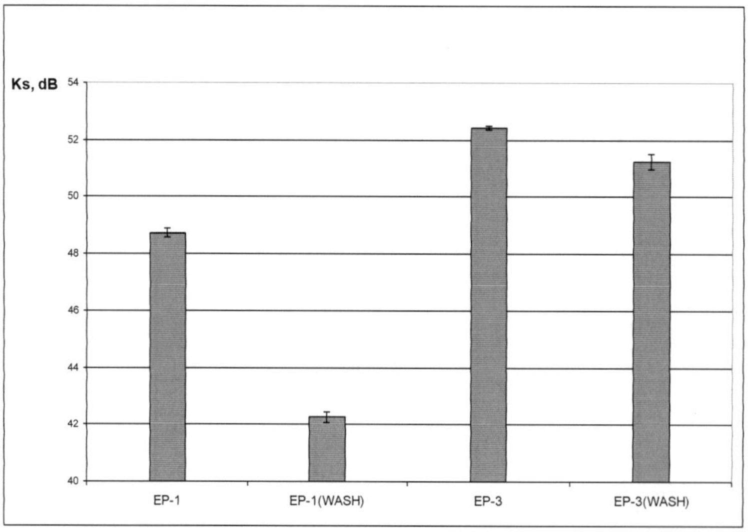


**Figure 1.** Experimental results of shielding coefficient values assessment for different IPM samples.

IOP Conf. Series: Materials Science and Engineering **487** (2019) 012031 doi:10.1088/1757-899X/487/1/012031

Inside EP-1 EF levels were from 0.027 to 0.303 kV/m, inside the EP-3 – from 0.017 to 0.202 kV/m. EF levels of washed EP-1 were from 0.058 to 0.63 kV/m, washed EP-3 – from 0.021 to 0.225 kV/m (after 10 washes). Inside EP-1 and EP-3 EF levels increased after IPM washing.

According to experimental results shown in figure 1, the considered IPM samples satisfy with current technical requirements, and its shielding factors are higher than minimum permissible value of 30 dB. The measured inside the IPM electric field levels are less than 1 kV/m and don't exceed International and Russian hygienic standards for occupational electromagnetic field exposure. As test measurements of electric field levels were performed only for one point in human body phantom it is important to estimate the IPM shielding efficiency for whole body volume. For this goal, the numerical simulation with human body phantom was used.

The developed numerical model of test setup was verified by means of the EF spatial distribution compared with experimental results of EF measurements inside air capacity. The IPM test condition models were based on the compatibility of these data.

IPM numerical model was exposed to of 70.62 kV/m EF level and provided shielding coefficient value of 50.13 dB. The averaged EF level inside IPM model was 220 mV/m. The selected values of external EF strength and shielding efficiency were almost equal to experimental data.

The assessment of EF levels induced in human body tissues was conducted for worker' body model exposed to 70.62 kV/m EF level and for worker' body dressed in IPM model exposed to the same level. The spatial distributions of internal EF for each case are shown in figure 2 in different scales.

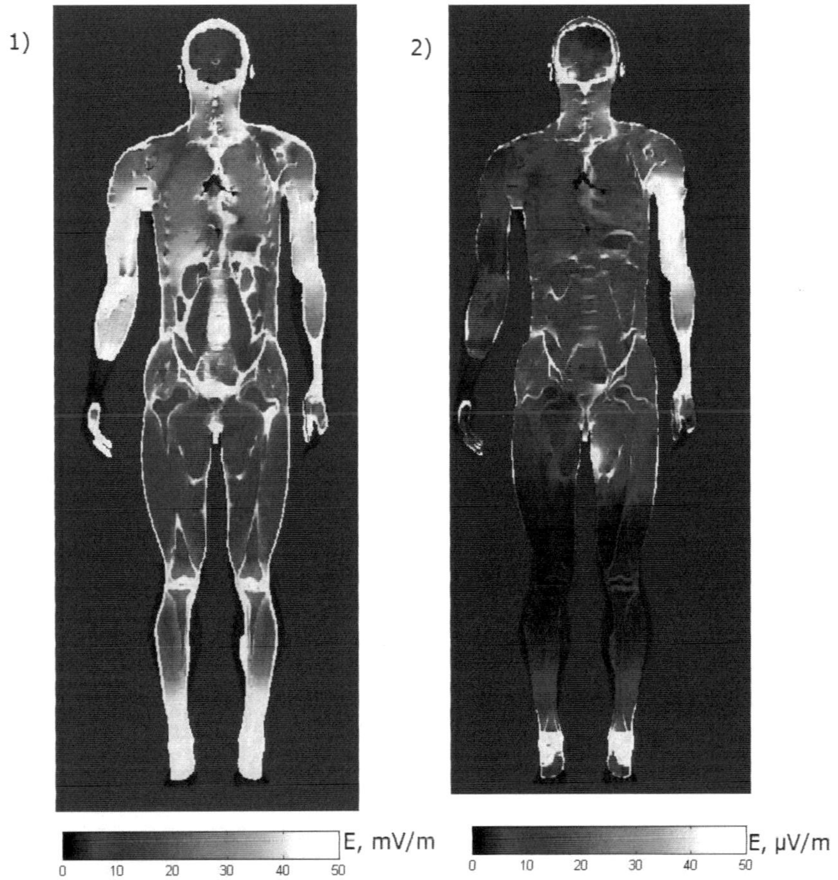


**Figure 2.** Spatial distributions of EF induced in human body tissues for cases: 1) – worker' body without IPM; 2) – worker' body dressed in IPM.

The simulation results show that the induced EF distributions inside human tissues in the case of IPM model (figure 2, 2) is high non-symmetric when compared with human body exposed non-shielded EF. This can be caused by IPM model inaccuracy and incomplete. However, the induced EF levels are reduced significant in the shielded field, see table 2.

**Table 2.** The numerical simulation results of induced electric field value in human body.

| Tissue | Induced electric field E (V/m) | |
|---|---|---|
| | Worker without PPM | Worker with PPM |
| Brain | 0.473 | 0.037 |
| Cerebellum | 0.081 | 0.029 |
| Spinal Cord | 0.217 | 0.008 |
| Heart | 0.160 | 0.003 |
| Muscle | 3.543 | 0.150 |
| Lungs | 0.159 | 0.004 |
| Liver | 0.128 | 0.005 |
| Spleen | 0.091 | 0.004 |

Presented in table 2 EF levels induced in several organs of human body exposed by high-intensity 50 Hz EF which elevated the international norms. For example, for brain and muscle these values are 4 times higher than ICNIRP [3] basic restriction levels and can cause injuries for worker. IPM use allows reducing the induced EF levels and ensures worker safety.

The shielding EF is one of the basic principles for worker protection against high intensity PF EF. The EF strength measurements are used for experimental assessment of IPM shielding efficiency and provide data compared with hygienic standards. The results of such tests can be insufficient and incomplete, and numerical simulation might be used as additional technique.

**References**

[1] *SanPiN 2.2.4.3359-16 Sanitary-Epidemiological Requirements for Physical Factors in the Workplace* (in Russian)
[2] Rubtsova N et al E 2018 *Electromagn. Biol. Med.* **37** 43–9
[3] 2010 *ICNIRP Guidelines for Limiting Exposure to Time-Varying Electric and Magnetic Fields (1 Hz to 100 kHz)* 2010 *Health Physics* **99** 818–36
*[4] Directive 2013/35/EU of 26 June 2013 on the Minimum Health and Safety Requirements Regarding the Exposure of Workers to the Risks Arising from Physical Agents (electromagnetic fields)*
[5] *TR TS 019/2011 On the Safety of Personal Protective Means* (in Russian)
[6] *GOST 12.4.172-2014 Set of Individual Shielding for Protection Against Electric Fields of Power Frequency. General Technical Requirements and Control Methods* (in Russian)
[7] Christ A et al 2010 *Phys. Med. Biol.* **55** 23–38

IOP Conf. Series: Materials Science and Engineering **487** (2019) 012032 doi:10.1088/1757-899X/487/1/012032

# Effect of acute UV irradiation of barley in different stages of organogenesis on yield

**P N Tsygvintsev, O A Guseva and M Yu Tatarova**

Russian Institute of Radiology and Agroecology, Kievskoe sh. 109 km, Obninsk, Russia

E-mail: paul-gomel@mail.ru

**Abstract**. The negative effect of acute barley irradiation depends both on the spectrum of UV radiation and on the stage of plant development. UV-B radiation, having a higher photon energy, has the greater negative effect compared to UV-A radiation at the same dose of irradiation. During the vegetation, barley is the most sensitive to the influence of ultraviolet at the II-V stages of organogenesis (in the transition from the vegetative to the generative stage of development). The adverse effect of UV-B radiation on the grain yield shows in the substantial reduction of flowers in the ear during their laying and probably in the pollen sterility.

## 1. Introduction

Ultraviolet (UV) radiation occupies a special place among global environmental factors affecting terrestrial ecosystems and, hence, the agrosphere. The decrease in the concentration of stratospheric ozone increases the solar UV radiation influx at the earth's surface and shifts it to a shorter wavelength spectral range. By the end of the 20th century, the reduction of the ozone concentration in the stratosphere over the countries of Europe was about 1% per year [1]. At the present stage of the biospheric phenomena, this tendency can be intensified due to appearance of the "ozone holes" in the winter-spring period, previously not observed in the Arctic. Sharp decreases in stratospheric ozone in the Arctic in spring attracted strong attention to studies of the ultraviolet radiation effect on agricultural plants [2]. In this connection, it became necessary to experimentally study the effects of acute UV irradiation of agricultural agrocenoses.

Particular attention should be given to the study of the sensitivity of plants at different stages of organogenesis, since an increase in the intensity of UV-B radiation due to the formation of "ozone holes" is usually observed in the spring period when the active growth of winter grain crops begins. As can be seen in figure 1, the ozone depletion can be as high as 50% in some cases. For plants in the northern part of Eurasia and Canada, the usual level of ultraviolet radiation is very low during this period because of the low sun position above the horizon. And plants may not have time to adapt to a sharp and short-term increase in UV radiation [3]. Research of the vegetating plants reaction to acute UV irradiation at various stages of organogenesis is necessary to predict crop losses. Therefore, this article is focused on the morphological reaction of barley to acute UV irradiation (A and B) in different phases of organogenesis.

IOP Conf. Series: Materials Science and Engineering **487** (2019) 012032 doi:10.1088/1757-899X/487/1/012032

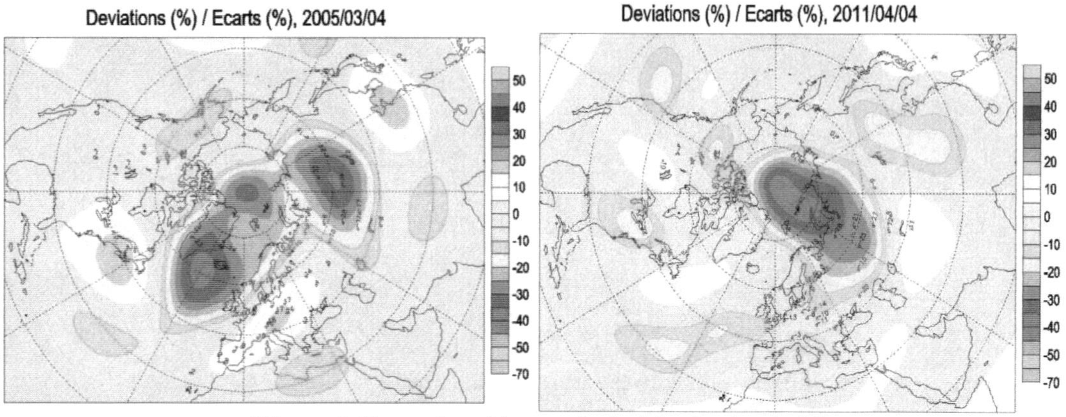


**Figure 1.** Examples of "ozone holes" in the Arctic [4].

## 2. Material and methods

The subject of the study was barley (Hordeum vulgare L.) variety Zazersky 85. The plants were grown in vessels containing 4.5 kg of air-dry sod-podzolic soil. Agrochemical properties of soil: $pH_{KCl}$ 5.00±0.01; humus 1.42±0.01%; hydrolytic acidity 2.06±0.03 meq / 100 g soil; the sum of the absorbed bases is 5.75±0.05 meq / 100 g soil; exchange $K_2O$ and mobile $P_2O_5$ 8.52±0.4 and 48.5±1.0 mg / 100 g, respectively. Prior to sowing, nutrients were added to the soil at the rate of $N_{200}$, $P_{100}$ and $K_{100}$ mg / kg of soil according to the active substance in the form of aqueous solutions of salts ($NH_4NO_3$, $KH_2PO_4$ and $K_2SO_4$).

The irradiation was carried out once when the plants reached certain phases of organogenesis: the third leaf (stage I of organogenesis), the tillering (stage II organogenesis), the outlet into the tube (stage III of organogenesis) and the formation of the lower node of the straw (stage V organogenesis) for 5 hours at a dose rate 5 $W/m^2$. The total dose was 90 $kJ/m^2$. As a source of UV-A radiation, Black Light BLUE lamps from Philips, for UV-B — LER-40 (figure 2) were used. Since a decrease in the thickness of the ozone layer to 50%, a significant increase in UV-B radiation is observed only in the range of 295 – 320 nm, and the greatest increase in UV-A in the 320 – 370 nm range (figure 3), in the first approximation can be considered irradiation with these lamps corresponding to the spectrum of acute irradiation due to "ozone holes".

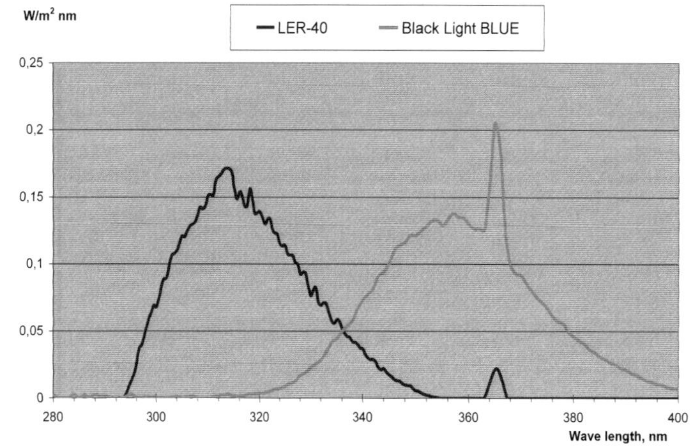


**Figure 2.** Spectra of LER-40 and Black Light BLUE lamps in the UV range.

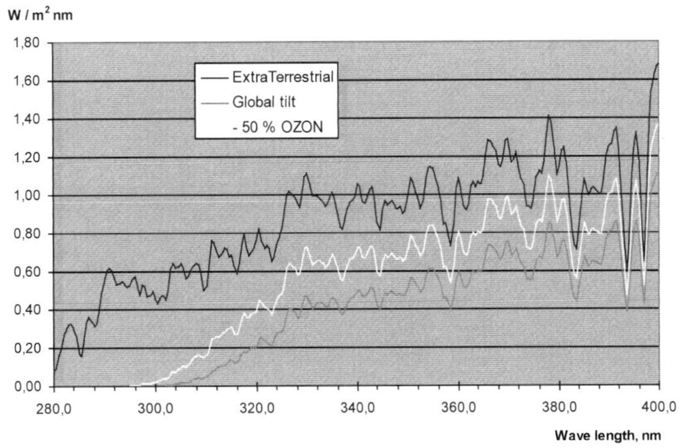


**Figure 3.** Spectra of solar radiation in the UV range behind the atmosphere, at ground level and at a 50% reduction in the ozone layer.

The effect of the studied spectra of acute UV radiation was estimated at the end of vegetation through the morphological parameters and the grain yield from the plant. The reliability of the differences in the variants of the experiment was established according to the t-test. The text and figures show the mean values and the error of the mean X ± m.

## 3. Results

The study of the morphological response of barley to acute irradiation of UV-A and UV-B ranges when exposed to various stages of organogenesis showed an ambiguous dependence.

UV irradiation at these stages of organogenesis did not have a negative effect on the barley growth (figure 4). UV-B irradiation at the V stage of organogenesis caused suppression of plant height by 12% ($p < 0.05$). The integral indicator that determines the resistance of plants to various stress factors is the final yield (figure 5), the decrease of which indicates the changes occurring at different levels of the organization of the plant organism and the violation of its most important functions.

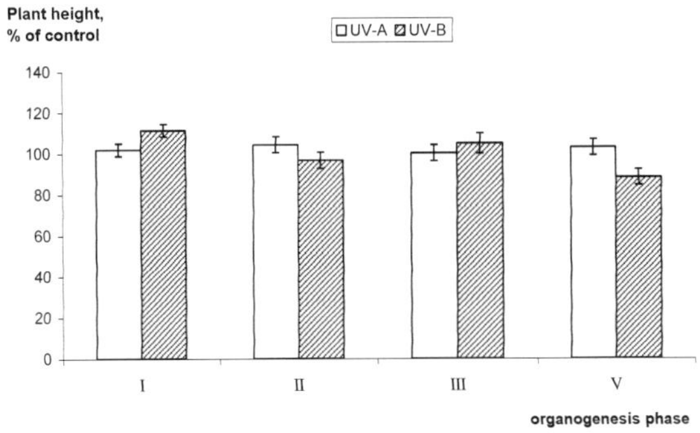


**Figure 4.** Plant height after acute UV irradiation in different phases of ontogenesis in relation to control.

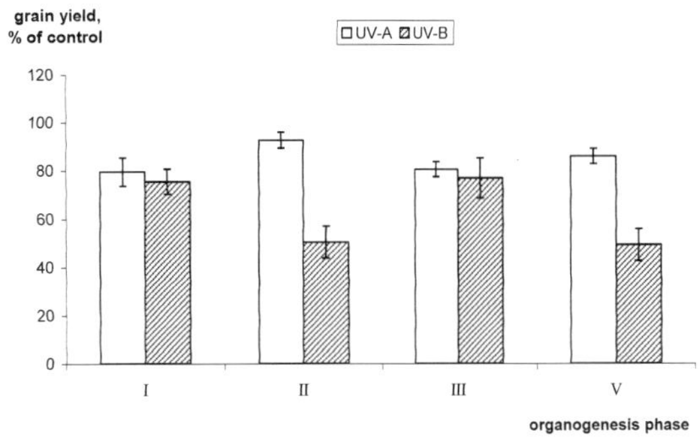


**Figure 5.** Grain yield after acute UV irradiation in different phases of ontogenesis in relation to control.

As can be seen in figure 5, the greatest yield losses occur when plants are irradiated with UV-B in the phase of formation of additional shoots and in the phase of flower formation. If we consider the effect of acute UV irradiation of barley plants in terms of the additional yield shoots formation on the plant, UV-A irradiation during the first growth period even contributes to their increase ($p<0.05$), while UV-B irradiation significantly reduces the productivity of additional stems in barley plants in any period of development (figure 6). The effect of chronic increased UV-A and UV-B irradiation on the formation of lateral shoots was noted in the other studies [5]. Acute UV irradiation also negatively affects the formation of grain in the ear of plants. In this case, the most pronounced effect is from UV-B irradiation in the phases when the ear (II) is laid and flowers (V) are formed (figure 7).

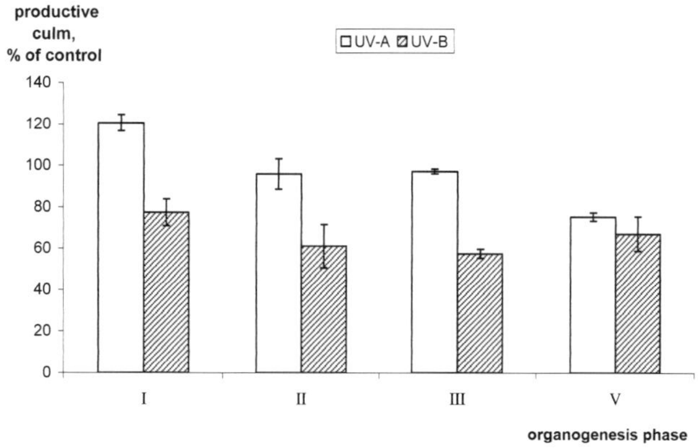


**Figure 6.** Number of productive shoots per plant after acute UV irradiation in different phases of ontogenesis in relation to control.

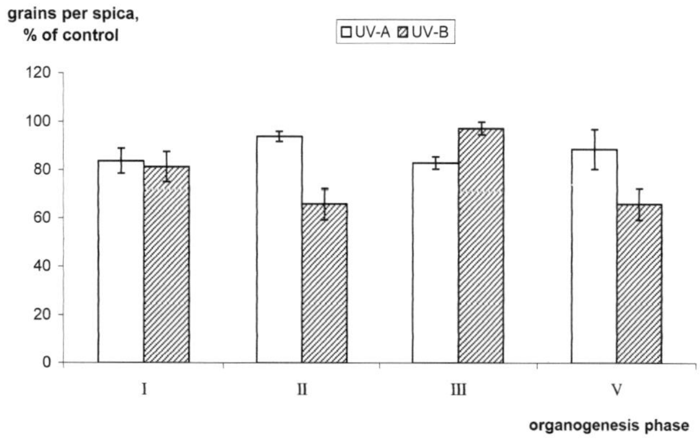


**Figure 7.** The number of grains in the ear after acute UV irradiation in different phases of ontogenesis in relation to the control.

Thus, as a result of the studies, it was revealed that the UV-B irradiation of plants has more negative influence then UV-A on all morphological parameters studied (plant height, number of productive stems, number of grain in the ear and mass of grain from the plant) during all stages of organogenesis at a total daily dose of 90 kJ/m$^2$. The negative effect of UV-B radiation was manifested in the formation of productive stems during irradiation at all stages of organogenesis (the decrease amounted to an average of 40%). UV-B irradiation of barley in the II and V stages of organogenesis led to a decrease in the average head spike of an average stem by 34% compared to the control, whereas UV-A irradiation showed no differences from the control for this indicator. UV-B irradiation in the II-V stages of organogenesis also caused a decrease in the mass of each grain by an average of 48%.

The obtained experimental data cannot be used directly to predict the loss of barley yield from acute ultraviolet irradiation, but only to evaluate the most sensitive phases of the development of barley plants in the ultraviolet. In this experiment, separate UV-A and UV-B irradiation was used. However, there are data on the "protective" effect of UV-A radiation relative to the effect of UV-B radiation upon their combined action [6]. It is also necessary to take into account the density of crops, since the effect in photobiology is tied to the flux density (J / $m^2$), and the directly absorbed dose (J / kg) can depend both on the stage of plant formation [7] and on their seeding density. Therefore, the task of estimating the loss of harvest of winter grain crops from acute UV radiation as a result of the formation of "ozone holes" over the Arctic requires further, more detailed, research.

**References**

[1] Caldwell M M, Ballare C L, Bornman J F, Flint S D, Bjorn L O, Teramura A H, Kulandaivelu G and Tevini M 2003 Terrestrial ecosystems increased solar radiation and interactions with other climatic factors *Photochem. Photobiol. Sci.* **2** 29–38

[2] Strzhizhovskiĭ A D 1999 Plant reaction to elevated ultraviolet irradiation: potential impacts of stratospheric ozone depletion *Radiation Biology. Radioecology* **39** 683–92

[3] Klem K, Holub P, Štroch M, et al 2015 Ultraviolet and photosynthetically active radiation can both induce photoprotective capacity allowing barley to overcome high radiation stress *Plant Physiology and Biochemistry* **93** 74–83

[4] Site data http://exp-studies.tor.ec.gc.ca/cgi-bin/selectMap

[5] Hakala K, Jauhiainen L, Koskela T, et al 2002 Sensitivity of crops to increased ultraviolet radiation in northern growing conditions. *Journal of Agronomy and Crop Science* 2002 **188** 8-18

[6] Štroch M., Materová Z., Vrábl D., et al 2015 Protective effect of UV-A radiation during acclimation of the photosynthetic apparatus to UV-B treatment *Plant Physiology and Biochemistry* **96** 90–6

[7] Tsygvintsev P N 2017 Absorption of UV by barley at different stages of organogenesis *Int. J. Applied and Fundamental Res.* **10** 283–9